Springer Theses

Recognizing Outstanding Ph.D. Research

Aims and Scope

The series "Springer Theses" brings together a selection of the very best Ph.D. theses from around the world and across the physical sciences. Nominated and endorsed by two recognized specialists, each published volume has been selected for its scientific excellence and the high impact of its contents for the pertinent field of research. For greater accessibility to non-specialists, the published versions include an extended introduction, as well as a foreword by the student's supervisor explaining the special relevance of the work for the field. As a whole, the series will provide a valuable resource both for newcomers to the research fields described, and for other scientists seeking detailed background information on special questions. Finally, it provides an accredited documentation of the valuable contributions made by today's younger generation of scientists.

Theses are accepted into the series by invited nomination only and must fulfill all of the following criteria

- They must be written in good English.
- The topic should fall within the confines of Chemistry, Physics, Earth Sciences, Engineering and related interdisciplinary fields such as Materials, Nanoscience, Chemical Engineering, Complex Systems and Biophysics.
- The work reported in the thesis must represent a significant scientific advance.
- If the thesis includes previously published material, permission to reproduce this must be gained from the respective copyright holder.
- They must have been examined and passed during the 12 months prior to nomination.
- Each thesis should include a foreword by the supervisor outlining the significance of its content.
- The theses should have a clearly defined structure including an introduction accessible to scientists not expert in that particular field.

More information about this series at http://www.springer.com/series/8790

Bayode Owolabi

Characterisation of Turbulent Duct Flows

Experiments and Direct Numerical Simulations

Doctoral Thesis accepted by
the University of Liverpool, Liverpool,
United Kingdom

 Springer

Author
Dr. Bayode Owolabi
Department of Mechanical Engineering
University of Alberta
Edmonton, AB, Canada

Supervisor
Dr. David J. C. Dennis
School of Engineering
University of Liverpool
Liverpool, UK

ISSN 2190-5053   ISSN 2190-5061   (electronic)
Springer Theses
ISBN 978-3-030-19747-6   ISBN 978-3-030-19745-2   (eBook)
https://doi.org/10.1007/978-3-030-19745-2

This Springer imprint is published by the registered company Springer Nature Switzerland AG
The registered company address is: Gewerbestrasse 11, 6330 Cham, Switzerland

Supervisor's Foreword

Turbulence is prevalent in the natural world (for example, the earth's atmosphere and the human respiratory system) as well as many important engineering flows. One of the most common engineering applications is that of the transport of fluids (for example, water, oil, gas or air) through pipes or ducts. This thesis presents several new findings in the field of turbulent duct flows, which are important for many industrial applications. The work significantly advances our understanding of the fundamental physics behind these flows. Unusually, it includes both high-quality experiments and cutting-edge numerical simulations, providing a level of insight and rigour rarely achievable by a Ph.D. student.

The scientific advancements include the revelation that the earth's rotation can change the laminar flow in a large duct to such an extent that it significantly affects the transition to turbulence. This is an example of a phenomenon that goes against the usual scientific assumptions (i.e. that the earth's rotation does not influence lab-scale experiments) to such an extent that even experienced researchers struggle to believe it until presented with unequivocal evidence like that contained in this thesis.

The thesis also includes a comprehensive experimental study of the phenomenon of drag reduction by polymer additives in turbulent duct flows. Whilst turbulent drag reduction by polymer additives has been recognised for decades, it is still not fully understood. In particular, the level of drag reduction achieved has not been reliably predictable. The experiments in this thesis provide a simple correlation that gives a quantitative prediction of drag reduction from a single, measurable material property of the polymer solution, independent of flow geometry and concentration. This is not something that has been achieved previously and is an important advancement from both a scientific and practical perspective.

In addition, this thesis includes the first experimental evidence of marginal turbulence occurring in a pressure-driven square duct flow, something that was only before predicted by simulations, as well as the identification of similar marginal turbulence in wall-driven flows using simulations (for the first time by any means). This marginally turbulent regime is an interesting discovery, which could prove very useful for flow control in the future. It is where the flow switches back and forth between two distinct flow states, which underlie, and are to some extent hidden by, the turbulence.

This thesis contains an excellent and varied body of work, which uses both experimental and numerical approaches to make several valuable contributions to the field of fluid mechanics. It is a fine achievement, well deserving of this recognition from Springer.

Liverpool, UK   Dr. David J. C. Dennis
March 2019

Abstract

Turbulent duct flows are encountered in a wide range of engineering applications; a fundamental physical understanding of such flows is thus very important for making predictions about heat transfer, mixing, and skin friction drag. Previous studies have focused mainly on the purely pressure driven (Poiseuille) case at relatively large Reynolds numbers (Re), hence the flow characteristics at low Re are not well understood. Limited Poiseuille flow direct numerical simulation (DNS) data show that in non-circular ducts, there exists an interesting phenomenon at Re close to transition to turbulence. Specifically in ducts of square cross-section, the flow field is observed to switch between two states characterised by different secondary flow patterns, thus potentially having serious implications for heat and mass transport. This finding has never been verified experimentally.

Another important area in the study of turbulent duct flows, which is yet to be fully understood, is the drag reduction (DR) obtained by seeding with long-chain flexible polymers having high molecular weights. These are known to modify the turbulence field in duct flows when added even at minute concentrations, causing a massive decrease in skin friction; yet it has never been possible to relate the degree of DR to a measurable fluid property.

In this study, turbulent duct flows of Newtonian and non-Newtonian fluids over a wide range of Reynolds numbers are investigated both experimentally and numerically. The geometries considered include a square duct, rectangular channel and circular pipe. In purely pressure-driven flow in a square duct, the onset criteria for transition to turbulence is first examined. In so doing, the potential importance of Coriolis effects on this process for low-Ekman-number flows is highlighted. Experimental data on the mean flow properties and turbulence statistics at relatively low Reynolds numbers are then obtained. The alternation of the flow field between two states in the "marginal turbulence" regime in a square duct, originally predicted by the DNS of Uhlmann et al. 2007, is confirmed by bimodal probability density functions of streamwise velocity at certain distances from the wall as well as joint probability density functions of streamwise and wall-normal velocities which feature two peaks highlighting the two states. By applying Taylor's hypothesis of frozen turbulence to the data, it is shown that there is also a spatial switching along the length of the duct.

Similarly, direct numerical simulations of zero-net-flux wall-driven (Couette) flow in a square duct reveal an alternation between two states, thus indicating that the phenomenon is not unique to Poiseuille flows. The secondary motions are observed to be closely related to the near-wall ejection and sweep events. Furthermore, the side walls are found to have a stabilising effect on the flow, the critical Reynolds number for transition being much higher than that in plane Couette flow.

For an experimental investigation of square duct Couette–Poiseuille flows, a new test section with one moving wall has been designed and constructed. Preliminary Laser Doppler Velocimetry (LDV) measurements of the velocity profiles in the laminar regime show that the fully developed analytical solution can be accurately reproduced in the facility. Suggestions for future turbulent flow studies in the new test section have been given.

Finally, the polymer DR problem has been revisited, and for the first time, a correlation which allows for quantitative predictions of DR from the knowledge of a single measurable material property of a polymer solution, independent of the geometry, concentration, and other experimental variables is obtained.

Acknowledgements

My utmost thanks go to God Almighty for giving me the wisdom, knowledge and understanding to successfully undertake the research presented in this thesis.

My profound gratitude also goes to my supervisors, Dr. David Dennis, Prof. Robert Poole and Prof. Chao-An Lin. Without their guidance, support and encouragement, this work would not have been possible. I have been greatly motivated by their passion and enthusiasm for research.

I am very grateful to the University of Liverpool and National Tsing Hua University Taiwan for granting me a scholarship to undertake this dual Ph.D. programme. I would also like to acknowledge the support from the Federal University of Technology Akure, Nigeria.

I would like to say a big thank you to my colleagues in Liverpool (both past and present): Mr. Allysson Domingues, Mr. Rishav Agrawal, Mr. Mahdi Davoodi, Mr. Oguzhan Der, Dr. Chaofan Wen, Dr. Waleed Abed, Dr. Henry Ng, Dr. Zografos Konstantinos, Dr. Haonan Xu and Dr. Simeng Chen. They have always been ready to offer assistance when needed. I am grateful to Mr. Xianke Lu for assisting in translating the abstract of this thesis from English to Chinese language. My thanks also go to colleagues at National Tsing Hua University for making my stay in Taiwan enjoyable. Specifically, I am grateful to Ms. Sung-Hua Chen and Mr. Che-Yu Lin for being very good friends.

I gratefully acknowledge the technical support provided by Mr. John Curran, Mr. Derek Neary and Mr. Christopher Hinchliffe during the course of this research and would also like to thank Mr. Jack Carter-Hallam and Ms. Chih-Hsuan Hsu for their excellent administrative support.

I would like to express my appreciation to Rev. Dr. Johnson Abimbola and my friends at Alive Believers Centre for making me feel at home in Liverpool. Finally, I sincerely appreciate the love and support of my parents Prof. and Mrs. Owolabi and my brothers John and David. Their words of encouragement have kept me going.

Thank you all.

Contents

Nomenclature

Roman Symbols

A	Cross-sectional area of duct (m^2)
B	Log-law constant
C_i	Convection term in Navier–Stokes equation (N/m^3)
C_1	Limiting value of $\%DR$ as $Wi \to \infty$
c	Concentration (ppm)
c^*	Critical overlap concentration (ppm)
c_{ij}	Conformation tensor
D	Duct diameter or height (m)
D_h	Hydraulic diameter (m)
D_i	Diffusion term in Navier–Stokes equation (N/m^3)
d_f	Fringe spacing (m)
D_{mid}	Mid diameter of filament (m)
D_0	Diameter of circular plates in CaBER (m)
De	Deborah number
E	Relative error in velocity computation (%)
e	Error in development length estimate (%)
Ek	Ekman number
f	Fanning friction factor
$f_{Blasius}$	Fanning friction factor from Blasius' correlation
f_D	Doppler shift (Hz)
$\mathcal{F}_i$	Transit time weighting
f_i	Coriolis force per unit mass (N/kg)
f_s	Frequency shift (Hz)
f_{Virk}	Fanning friction factor from the correlation of Virk (1975)
G	Elastic modulus (Pa)
H	Channel height (m)
h	Half height of duct (m)

I	State indicator function
II	Second invariant of Reynolds stress anisotropy tensor
III	Third invariant of Reynolds stress anisotropy tensor
K	Consistency (Pa.s^n)
K_{cy}	Constant in Carreau–Yasuda model (s)
L	Development length (m)
$\mathcal{L}$	Maximum extension of strings in FENE-P model (m)
ℓ	Length over which pressure drop is measured (m)
l	Mixing length (m)
L_c	Critical length of duct for self-sustained turbulence to exist (m)
L_{fluent}	Development length from ANSYS Fluent (m)
L_{extrap}	Development length estimate from Richardson extrapolation (m)
L_x	Length of a duct (m)
m	Parameter in Carreau–Yasuda model introduced by Yasuda et al. (1981)
N	Number of samples
n	Power law index or time step (Chap. 4)
N_x, N_y, N_z	Number of grid cells in the x, y and z directions
p	Instantaneous pressure (Pa)
$\bar{p}$	Mean pressure (Pa)
q	Wetted perimeter of duct (m)
R	Radius (m)
r	Ratio of wall to bulk velocity
r_0	Radial distance (m)
$R_{u'u'}$	Streamwise velocity autocorrelation function
Re	Reynolds number based on bulk velocity and half height or radius
Re_c	Critical Reynolds number for transition to turbulence
Re_δ	Reynolds number based on hydraulic diameter
Re_τ	Reynolds number based on friction velocity. $Re_\tau = u_\tau h/\nu$
Re_w	Reynolds number based on wall velocity. $Re_w = U_w h/\nu$
S	Skewness
$\mathcal{S}$	Magnitude of secondary flow (m/s)
s	Azimuthal distance (m)
$\mathcal{T}$	Torque (N.m)
t	Time (s)
t_i	Transit time of a seeding particle through a measurement volume (s)
T_0	Characteristic time of a process (s)
Tr	Trouton ratio
u, v, w	Instantaneous streamwise, transverse and spanwise velocity (m/s)
$\tilde{u}$	Instantaneous streamwise, velocity averaged over the length of a duct (m/s)
$\bar{u}, \bar{v}, \bar{w}$	Mean streamwise, transverse and spanwise velocity (m/s)
U_b	Bulk velocity (m/s)

U_c	Streamwise centreline velocity (m/s)
$\mathcal{U}_c$	Convection velocity in the application of Taylor's hypothesis (m/s)
u_i	Velocity field (m/s)
$\overline{u_i'u_j'}$	Reynolds stress tensor (m^2/s^2)
$\overline{u'^2}, \overline{v'^2}, \overline{w'^2}$	Reynolds normal stresses (m^2/s^2)
$\overline{u'v'}, \overline{v'w'}, \overline{u'w'}$	Reynolds shear stresses (m^2/s^2)
u_{max}	Maximum streamwise velocity (m/s)
$u_{max,a}$	Fully developed analytical value of u_{max} (m/s)
u_{rms}	Root mean square streamwise velocity fluctuation (m/s)
u_τ	Friction velocity, $\sqrt{\overline{\tau_w}/\rho}$ (m/s)
u_τ^*	Friction velocity at the centre of a wall (m/s)
U_w	Wall velocity (m/s)
V	Velocity component perpendicular to fringe pattern (m/s)
$\mathcal{V}$	Volume (m^3)
v_{rms}	Root mean square transverse velocity fluctuation (m/s)
W	Channel width (m)
Wi	Weissenberg number
Wi_c	Critical Weissenberg number
x_i	Spatial co-ordinate (m)
x, y, z	Streamwise, transverse and spanwise co-ordinates (m)

Greek Symbols

α	Angle between duct axis and rotation axis of the earth ($^\circ$)
α_L	Latitude ($^\circ$)
α_N	Angle between the direction of true north and duct axis ($^\circ$)
β	Solvent to total viscosity ratio
γ	Degree of compression of a grid near a boundary
$\dot{\gamma}$	Shear rate (s^{-1})
δ_{ij}	Kronecker delta
$\Delta x, \Delta y, \Delta z$	Grid spacing in x, y, z directions (m)
$\overline{\Delta p}$	Mean pressure drop (Pa)
$\overline{\Delta p_N}, \overline{\Delta p_V}$	Mean pressure drop in Newtonian and viscoelastic fluids (Pa)
$\dot{\varepsilon}$	Strain rate (s^{-1})
ϵ	Dissipation rate per unit mass (m^2/s^3)
ζ, χ	Co-ordinates of a uniformly spaced mesh (m)
η	Kolmogorov length scale (m)
$[\eta]$	Intrinsic viscosity (ml/g)
η_{sp}	Specific viscosity
θ	Cone angle ($^\circ$)
κ	Von Karman constant
Λ	Wavelength of laser beam (nm)

λ	Relaxation time (s)
λ_c	Polymer relaxation time measured using CaBER (s)
μ	Dynamic viscosity (Pa.s)
μ_0	Zero shear rate viscosity (Pa.s)
μ_∞	Infinite shear rate viscosity (Pa.s)
ν	Kinematic viscosity, μ/ρ (m^2/s)
ξ	Parameter which controls the symmetry of a grid
ρ	Density (kg/m^3)
σ	Surface tension (N/m)
$\bar{\tau}$	Mean local wall shear stress (Pa)
$\bar{\tau}^*$	Mean shear stress at the centre of a wall (Pa)
τ_{ij}	Stress tensor (Pa)
$\bar{\tau}_w$	Wall shear stress averaged over the perimeter of a duct (Pa)
τ_{xy}	Shear stress (Pa)
$\bar{\tau}_{xy}$	Mean shear stress (Pa)
$\overline{\tau_{xy}^p}$	Mean polymer stress (Pa)
Φ	Log indicator function
φ	Angle between laser beams ($^\circ$)
Ω	Angular velocity of the earth (rad/sec)
ω	Angular velocity (rad/sec)
$\bar{\Omega}_x$	Mean streamwise vorticity (s^{-1})

Superscripts

$+$	Normalisation by friction velocity or ν/u_τ
$'$	Fluctuating component of velocity
p	Polymer
$*, **$	Intermediate time steps

Subscripts

E, W, N	Centroids of neighbouring control volumes (Chap. 4)
S, T, B	Centroids of neighbouring control volumes (Chap. 4)
P	Centroid of a control volume (Chap. 4)
e, w, n, s, t, b	Faces of a control volume (Chap. 4)

Acronyms

CaBER	Capillary breakup extensional rheometer
CFL	Courant–Friedrichs–Lewy
DNS	Direct numerical simulation
DR	Drag reduction
HDR	High drag reduction
LDV	Laser Doppler velocimetry

LES	Large eddy simulation
PAA	Polyacrylamide: FloPAM AN934SH
PIV	Particle image velocimetry
RANS	Reynolds averaged Navier–Stokes
SPIV	Stereoscopic particle image velocimetry

List of Figures

List of Tables

Chapter 1
Introduction

1.1 Background

Turbulence is ubiquitous in nature and in engineering applications, occurring in a wide variety of fluid motions, ranging from the flow of water in a faucet to the motion of gasses in the photosphere of the sun; yet the phenomenon remains one of the least understood problems in classical physics and for centuries has continued to intrigue some of the world's best minds. The following quote is attributed to the famous British physicist Sir Horace Lamb:

> I am an old man now, and when I die and go to heaven, there are two matters on which I hope for enlightenment. One is quantum electrodynamics, and the other is the turbulent motion of fluids. And about the former I am rather optimistic [1].

So, what exactly is turbulence? To answer this question, it is useful to examine the characteristics of turbulent flows (see e.g. [2]). These include irregularity and high diffusivity (which results in rapid mixing). Furthermore, these flows are inherently time dependent and three dimensional. They are also highly dissipative, and thus bring about large energy losses. Turbulence is not a fluid property; rather, it is a feature of fluid flows and a continuum phenomenon that can be described by the conservation equations of mass, momentum and energy.

The challenge in turbulence research stems from the fact that the fluctuating motions cover a wide spectrum of spatial and temporal scales (the smallest scales, the Kolmogorov microscales, are set by viscosity, while the largest scales depend on the geometry). The governing mathematical equations are also highly non-linear, making them almost intractable.[1]

A category of turbulent motions, which over the years, has received a lot of attention from researchers, is that which occurs as a result of the interaction between a fluid and a solid boundary. Wall bounded turbulent flows are of immense engineering importance, and a large proportion of these occur in ducts. Examples include flows in

---

[1]Analytical solutions to the Navier-Stokes equations only exist for a few simple cases such as the fully-developed laminar flow in a circular pipe or between two infinite parallel plates.

© Springer Nature Switzerland AG 2019
B. Owolabi, *Characterisation of Turbulent Duct Flows*,
Springer Theses, https://doi.org/10.1007/978-3-030-19745-2_1

pipelines, heating, ventillation and air conditioning (HVAC) systems, nuclear reactor channels and jet engines, among others. In these applications, turbulence can produce both desirable and undesirable effects. Skin friction drag in pipelines, induced by turbulent motion can be responsible for up to fifty percent of the energy loss [3]. On the other hand, heating, ventilation and air conditioning systems rely on turbulence for efficient mixing and heat transfer. A fundamental physical insight into such flows is, therefore, crucial to ensuring improved efficiency in practical applications.

Historically, progress in understanding duct flows can be attributed to the interplay between experiments and theory, and in recent times, numerical simulations. Equipped only with basic tools, early researchers such as the French physicist and mathematician, Jean Poiseuille, and the French engineer, Henry Darcy studied the gross flow features and provided empirical relations for the pressure loss due to friction. Similarly, the renowned British professor, Osborne Reynolds investigated the transition from laminar to a turbulent flow in pipes. In his 1883 paper titled "An experimental investigation of the circumstances which determine whether the motion of water shall be direct or sinuous, and the law of resistance in parallel channels" [4], he showed the importance of a dimensionless number, now called the Reynolds number (named after him), which gives a measure of the relative contributions of inertia and viscous forces to the flow dynamics. As the Reynolds number is increased, inertial forces become more dominant, and a laminar flow eventually becomes turbulent. In another publication, Reynolds [5] suggested that the complex motions in the turbulent regime can best be described in terms of the average flow properties, thus paving the way for the statistical approach to turbulence research and the development of the Reynolds averaged Navier-Stokes (RANS) equations.

Later workers obtained detailed velocity field data as new measurement techniques became available. Although the methods were intrusive (introducing some disturbance into the flow), they nonetheless allowed for some useful insight into the turbulence structure to be gleaned. Nikuradse [6], under the supervision of Prandtl, conducted pitot tube measurements in non-circular ducts at relatively large Reynolds numbers and discovered the existence of secondary flows—mean motions in the ducts' cross-sections, which were superimposed on the main streamwise flow. Similarly, Laufer [7], using hot-wire anemometry, obtained turbulence data in a channel. Over the past fifty years however, optical measurement techniques such as laser Doppler velocimetry (LDV) and particle image velocimetry (PIV), which are less intrusive have been developed. Therefore, past experiments need to be revisited and new innovative studies contrived, to take advantage of the high spatial and temporal resolution now on offer, and the ability to obtain instantaneous velocity data across an entire cross-section or volume of flow.

While advanced measurement technologies were being developed for the study of complex flows, progress was also being made in the computation of these flows. In the RANS approach, flow quantities such as velocity and pressure are expressed as a superposition of their time-averaged values and a component which fluctuates about the mean. The Navier-Stokes equation is then averaged over time and an expression in terms of the mean flow quantities is obtained. A drawback of the RANS method is that additional terms—Reynolds (turbulent) stresses—are introduced, hence a situation in which there are more unknowns than equations arises; therefore, the turbulent stresses have to be modelled to close the equations. A vast amount of the literature is devoted to the development of turbulence models (see Nallasamy [8] for a detailed review). Usually, a number of assumptions based on intuition and dimensional analysis are made, hence RANS fails to provide a detailed representation of the flow. To obtain a highly accurate description of a turbulent flow field, every eddy, from the largest to the smallest must be accounted for, without incorporating any models. Therefore, in solving the governing equations numerically, the grid sizes and time steps must be small enough to resolve the Kolmogorov scales. This approach, referred to as direct numerical simulation (DNS) is, thus, very computationally expensive, hence only simple geometries at relatively low Reynolds numbers have, so far, been tackled by DNS [9]. Interestingly, at these low Reynolds numbers, new phenomena not previously observed by experiments have been discovered. For example, the fairly recent DNS of Uhlmann et al. [10] shows an alternation of the turbulence field in a square duct between two states, characterised by secondary flow patterns which are very different from what is observed in measurements at higher Reynolds numbers; but no experimental validation of these results exist.

Besides obtaining a good understanding of a turbulence field, another objective of turbulence research has been to develop various control techniques to either delay laminar/turbulent transition or modify the turbulence structures near the wall. Options that have been explored include surface modification (e.g. through the use of compliant coatings or by riblets—small streamwise-aligned protusions), wall oscillation, use of large eddy breakup (LEBU) devices, blowing and suction, introduction of body forces and addition of polymers among others. Polymers are especially known to produce dramatic effects in a turbulent flow [11]. When added to a fluid even at minute concentrations of a few parts per million, they bring about a massive decrease in the skin friction drag; hence they are of tremendous potential importance in applications which involve fluid transport in ducts. Although the phenomenon was discovered more than seven decades ago, the underlying mechanism is yet to be fully understood; therefore, there is need for continued research to obtain new insights on polymer drag reduction (DR).

In this study, turbulent duct flows of Newtonian and non-Newtonian fluids over a wide range of Reynolds numbers are examined using a combination of non-intrusive measurement techniques and direct numerical simulations. It is believed that experiments, numerical simulations and theory will continue to play important roles in improving our understanding of these flows.

1.2   Motivation for Current Study

Although significant progress has been made on the characterisation of wall-bounded turbulent flows, a number of issues remain unresolved. For example, despite the 135 years of research since the pipe flow study of Reynolds [4] was published, the mechanism for laminar/turbulent transition in certain flows is still not yet fully understood; importantly, the critical Reynolds number for the onset of self-sustained turbulence in ducts is still being debated.

Furthermore, at Reynolds numbers close to transition (so-called "marginal turbulence" regime), interesting observations have been made by DNS which are yet to be verified by experiments. As indicated in Sect. 1.1, the DNS of Uhlmann et al. [10] on pressure-driven flow in a square duct revealed a temporal switching of the flow between two states, at these Reynolds numbers. The secondary flow pattern in each state was shown to be characterised by four vortices rather than the conventional eight observed at large Reynolds numbers; hence, this could potentially have a significant impact on heat and mass transport. In this study, LDV measurements in a square duct will be conducted in the "marginal turbulence" regime to verify the DNS findings. It is also of interest to determine whether the alternation between different states is a ubiquitous feature of wall-bounded turbulent flows, as this is presently not clear; hence, DNS of marginally-turbulent wall-driven flow in a square duct will be carried out.

Finally, the polymer drag reduction problem will be revisited. Although there has been a large number of studies on this phenomenon, it has never been possible to relate the degree of DR to a measurable fluid property. This is due, in part, to the difficulty in accurately measuring these properties (especially the extensional rheological properties), and to the fact that polymer molecules degrade when subjected to high levels of deformation. With the introduction of the capillary breakup extensional rheometer (CaBER) [12, 13], it is now possible to obtain accurate extensional rheology data in dilute polymer solutions. Therefore, in this study, an attempt will be made to exploit the mechanical degradation of polymers in order to vary their rheological properties. A correlation which relates these properties to the degree of DR, independent of flow geometry, polymer concentration, degradation and other experimental variables will then be developed.

1.3   Aim and Objectives of Research

This research is aimed at investigating the turbulence characteristics in pressure and wall-driven duct flows. The focus is on Newtonian flows in a square duct at relatively low Reynolds numbers, but flow of dilute polymer solutions will also be examined, and comparative studies in a rectangular channel and cylindrical pipe undertaken. It is intended that a better insight into the dynamics of wall-bounded turbulence will be obtained.

The specific objectives of the study are as follows:

- to obtain experimental data on the turbulence field in pressure-driven (Poiseuille) square duct flows at relatively low Reynolds numbers in order to validate the DNS findings of Uhlmann et al. [10] on the alternation of the flow fields between two states;

- to conduct direct numerical simulations of wall-driven (Couette) turbulent flows in a square duct at relatively low Reynolds numbers in order to determine whether the switching between two states is unique to pressure-driven flows;
- to design and construct a new experimental test section for investigating turbulent Couette-Poiseuille flows, in order to provide more insight into the effect of wall motion on the turbulence field in a duct;
- to obtain measurements of the velocity field and fluid rheological properties in turbulent polymer flows through ducts of circular, rectangular and square cross-sections over a wide range of Reynolds numbers; the goal is to develop a universal correlation which relates the degree of drag reduction to a single measurable extensional rheological property of a polymer solution.

1.4 Outline of Thesis

The work presented in this thesis is organised into eight chapters. In the current chapter, a brief background to the study has been given. The remainder of the thesis is arranged as follows.

In Chap. 2, an introduction to the theoretical concepts that will be applied in this thesis is given. Furthermore, a review of experimental and numerical studies on the pressure-driven as well as wall-driven turbulent flows of Newtonian fluids in non-circular ducts is presented. Previous findings on transition to turbulence and the existence of marginally turbulent states are also discussed. The chapter concludes with a discussion of previous works on turbulent duct flows with polymer additives and the polymer drag reduction phenomena.

In Chap. 3, the experimental arrangements and instrumentation used in this study are described in detail. The working principles of laser Doppler velocimetry and particle image velocimetry—two optical velocity measurement techniques employed in this research—are given. There is also a discussion on the techniques for the rheological characterisation of the working fluids and typical rheology data are presented.

A detailed description of the methods employed in the direct numerical simulation (DNS) of purely wall-driven Newtonian flows in a square duct is given in Chap. 4. Issues addressed include mesh generation, spatial and temporal discretisation of the governing transport equations, implementation of boundary conditions and validation of DNS code.

Chapters 5–7 contain the main results of this thesis. In Chap. 5, turbulent pressure-driven flows in a square duct at low Reynolds numbers are investigated. First, the onset criteria for transition to turbulence is determined and the potential importance of Coriolis effects on the process is highlighted. Data on the turbulence statistics and mean flow properties, obtained using LDV are then presented.

The DNS results on purely wall-driven turbulent flow in square duct are presented in Chap. 6. An investigation into the critical Reynolds number for a turbulent state in this flow is conducted. The secondary flow structure in the turbulent regime is examined in detail and a discussion on the flow statistics and turbulence structure is given.

The outcome of experiments on duct flows with polymer additives are discussed in Chap. 7. The drag reduction (DR) mechanism in a cylindrical pipe, a rectangular channel and a square duct—each a commonly encountered geometry in the study of wall bounded turbulence—is investigated. Specifically, the relationship between fluid elasticity and drag reduction is explored. Time-resolved velocity measurements using LDV and PIV at various levels of DR are also obtained.

Finally, in Chap. 8, the main conclusions from the research presented in this thesis are highlighted and recommendations for further work, provided. In the appendix, preliminary experimental results, restricted to laminar flow, from a newly designed test-section for the investigation of wall-driven flows in a square duct are presented. The results of numerical simulations on flow development length in the geometry are also given.

References

1. Lamb C (1932) Address to the british association for the advancement of science. Quoted in computational fluid mechanics and heat transfer by JC Tannehill, DA Anderson, and RH Pletcher 1984
2. Tennekes H, Lumley JL (1972) A first course in turbulence. MIT press
3. Marusic I, Mathis R, Hutchins N (2010) Predictive model for wall-bounded turbulent flow. Science 329(5988):193–196
4. Reynolds O (1883) An experimental investigation of the circumstances which determine whether the motion of water shall be direct or sinnous, and of the law of resistance in parallel channels. Philos Trans R Soc Lond A 174:935–82
5. Reynolds O (1895) On the dynamical theory of incompressible viscous fluids and the determination of the criterion. Philos Trans R Soc Lond A 186:123–164
6. Nikuradse J (1926) Untersuchungen über die geschwindigkeitsverteilung in turbulenten strömungen. VDI Forsch 281
7. Laufer J (1951) Investigation of turbulent flow in a two-dimensional channel. Technical Report 2123 N.A.C.A
8. Nallasamy M (1987) Turbulence models and their applications to the prediction of internal flows: a review. Comput Fluids 15(2):151–194
9. Jiménez J, Kawahara G (2012) Dynamics of wall-bounded turbulence. Cambridge University Press, pp 221–268
10. Uhlmann M, Pinelli A, Kawahara G, Sekimoto A (2007) Marginally turbulent flow in a square duct. J Fluid Mech 588:153–162
11. Toms BA (1948) Some observations on the flow of linear polymer solutions through straight tubes at large Reynolds numbers. In Proceedings of the 1st International Congress on Rheology, vol 2, pp 135–141
12. Bazilevsky A, Entov V, Rozhkov A (1990) Liquid filament microrheometer and some of its applications. In Third European Rheology Conference and Golden Jubilee Meeting of the British Society of Rheology. Springer, pp 41–43
13. Rodd L, Scott TP, Cooper-White JJ, McKinley GH (2005) Capillary break-up rheometry of low-viscosity elastic fluids. Appl Rheol 15:12–27

Chapter 2
Literature Review and Background Theory

2.1 The Mean Flow Profile and Law of the Wall in Turbulent Duct Flows

One major finding in fluid dynamics is the concept of boundary layers. Ludwig Prandtl in 1904 showed that the flow over a solid surface can be divided into two regions: a thin layer adjacent to the wall (the boundary layer), where the effects of viscous forces are important, and an outer irrotational flow region where these effects are negligible. In duct flows, the thickness of the boundary layers increase in the flow direction, eventually occupying the entire duct when the flow becomes fully developed. Close to a smooth surface the mean streamwise velocity ($\bar{u}$) can be assumed to only be a function of the flow conditions near the wall and independent of conditions elsewhere in the flow (see, e.g. [1]). Hence the only important parameters are the distance (y) from the wall, the mean shear stress at the wall ($\bar{\tau}_w$), fluid density (ρ), and kinematic viscosity ($\nu = \mu/\rho$, where μ is the dynamic viscosity). Dimensional analysis yields

$$\frac{\bar{u}}{u_\tau} = f\left(\frac{yu_\tau}{\nu}\right), \tag{2.1}$$

where $u_\tau = \sqrt{\bar{\tau}_w/\rho}$ is known as the friction velocity. Equation 2.1 is referred to as the "law of the wall". Near the wall, turbulent shear stresses are negligible, hence an approximately laminar shear flow exists. The total stress is given by

$$\bar{\tau}_w = \mu\frac{d\bar{u}}{dy}. \tag{2.2}$$

Integrating Eq. 2.2 and expressing the result in the same form as 2.1, yields the following linear expression for the mean velocity

$$\frac{\bar{u}}{u_\tau} = \frac{yu_\tau}{\nu}. \tag{2.3}$$

© Springer Nature Switzerland AG 2019
B. Owolabi, *Characterisation of Turbulent Duct Flows*,
Springer Theses, https://doi.org/10.1007/978-3-030-19745-2_2

Fig. 2.1 Mean velocity
profiles in a pipe (black line),
channel (red line) and
boundary layer (blue line).
$\kappa = 0.41$ and $B = 5.2$.
Adapted from El Khoury et
al. [4]

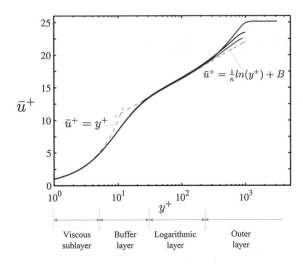

The region where Eq. 2.3 is valid is known as the viscous/laminar sublayer. Hereafter, $\bar{u}/u_\tau$ and yu_τ/ν are written as $\bar{u}^+$ and y^+, respectively.

Further from the wall, turbulent fluctuations become dominant. Applying Prantdl's mixing length concept gives

$$\bar{\tau}_w \approx \rho l^2 \left(\frac{\partial \bar{u}}{\partial y} \right)^2, \tag{2.4}$$

where l is the mixing length. Assuming that $l = \kappa y$, where κ is a constant, Eq. 2.4 integrates to the following expression, referred to as the logarithmic law

$$\bar{u}^+ = \frac{1}{\kappa} ln(y^+) + B, \tag{2.5}$$

where B is a constant. Over the years, the exact value of κ has been debated. Value ranges from 0.384 to 0.421 have been obtained from experimental data (see e.g. [2, 3]). A log indicator function,

$$\Phi = y^+ \frac{d\bar{u}^+}{dy^+}, \tag{2.6}$$

can be defined, whose constancy when plotted against y^+ confirms the presence of a truly logarithmic velocity distribution.

Figure 2.1 shows a typical turbulent velocity profile. The viscous sublayer extends from the wall to $y^+ \approx 5$, while the log-law is valid for $y^+ > 30$, and a buffer layer exists between the two. The region further away from the boundary, where the law of the wall ceases to hold is referred to as the outer/core flow region.

As will be shown in the following sections, turbulent flows in ducts with curved axes or straight ducts having non-circular cross-sections are also characterised by the existence of a mean secondary motion in a plane normal to the main (primary) flow direction. The magnitude of the secondary flow, $\sqrt{\bar{v}^2 + \bar{w}^2}$ (given that $\bar{v}$ and $\bar{w}$ are the mean velocity components in the cross-sectional plane and $\bar{u}$, the streamwise component, with y, z and x being the corresponding spatial coordinates), is usually small compared to the primary flow, $\bar{u}$ but its size relative to $\bar{u}$ depends on the driving physical mechanism. For a detailed discussion on the subject, the reader is referred to the publications of Johnston [5] and Bradshaw [6].

2.2 Pressure-Driven Flow in Non-circular Ducts

The earliest studies on pressure-driven flows in non-circular ducts can be traced back to the 1920s, when Nikuradse [7], under the supervision of Ludwig Prandtl, carried out pitot tube measurements of the axial velocity distribution in air flow in ducts of rectangular, square and triangular cross-sections, and observed a bulging of the contours of mean streamwise velocity towards the ducts' corners. Prandtl [8] attributed this phenomenon, which came to be named after him (Prandtl's secondary flow of the second kind) to the existence of secondary currents in the transverse plane, which transported momentum between the centre of a duct and the corners (See Fig. 2.2 for typical examples). These cross-stream motions are now known to be generated by gradients of Reynolds stresses, $\partial/\partial x_i (-\overline{\rho u_i' u_j'})$, in wall-bounded flows with no spanwise homogeneity [5, 6], hence they only exist in turbulent flows (since the Reynolds stresses vanish in a laminar flow). A second mechanism for the generation of secondary motions is the effect of lateral curvature of a main flow (in a curved duct for example [5]). A pressure gradient, $\nabla p = [\partial p/\partial y]\hat{i}_y + [\partial p/\partial z]\hat{i}_z$ is induced in the cross-sectional plane, in a direction pointing outward from the centre of curvature, thus setting up a secondary flow in the plane (see Fig. 2.3 for an illustration). Cross-stream motions generated by this mechanism are referred to as

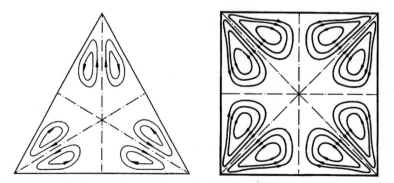

Fig. 2.2 Secondary flows in triangular and square channels. *Source* Stanković et al. [9]

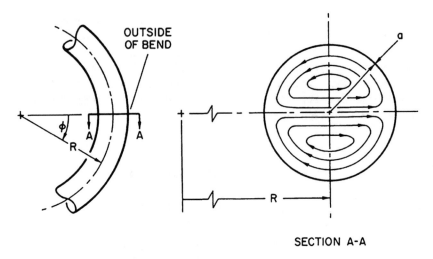

OUTSIDE
OF BEND

SECTION A-A

Fig. 2.3 Secondary flow streamlines in a curved pipe. *Source* Johnston [5]

Prandtl's secondary flow of the first kind and can occur in both laminar and turbulent flows.

Although the existence of Prandtl's secondary flow of the second kind was confirmed by the flow visualisation studies of Nikuradse [7], it was not until three decades later, with the development of the hot-wire anemometer, that an estimate of their magnitude was obtained [10]. Other flow measurement techniques such as laser Doppler velocimetry (LDV) and particle image velocimetry (PIV), which are considered to be non-intrusive, have since been developed [11, 12], thus making it possible to obtain accurate velocity measurements without disturbing a flow. Similarly, developments in computational fluid dynamics have made possible the large eddy and direct numerical simulations (LES and DNS) of turbulent duct flows and detailed turbulence statistics not easily obtainable in the laboratory are now available.

Turbulent flow in a square duct is perhaps the simplest flow exhibiting Prandtl's secondary motions of the second kind, making it ideal for fundamental studies on the structure of turbulent flows with mean three-dimensional velocity distribution, hence most of the research in the literature have been carried out in this geometry. In the following sections, a detailed review of both experimental and numerical studies on turbulent flow in a square duct is presented.

2.2.1   Turbulent Flow Experiments in a Square Duct

Following the findings of Nikuradse [7] on the existence of secondary currents in square duct turbulent flow, further experimental studies mostly at relatively large Reynolds numbers ($Re > 20000$; $Re = U_b h/\nu$, where h is the half height of a duct) have been conducted to obtain a better understanding of the phenomenon. The first quantitative study of secondary flows was carried out by Hoagland [10], who conducted detailed measurements of the fully developed streamwise and cross-stream velocity distributions in square and rectangular ducts (of aspect ratios 2:1 and 3:1) using a hot wire anemometer and pitot tube. The observed flow pattern was found to agree with Prandtl's prediction of a primary flow parallel to the duct's axis and a superimposed circulating flow in the cross-sectional plane, characterised by eight vortices, with symmetry about the corner and wall bisectors. The highest secondary velocities which were about 1.5% of the axial velocity along the duct centreline were found to occur close to the corners. From qualitative observations of the turbulence intensities, the author arrived at the conclusion that the secondary flow was in some way induced by turbulent fluctuations in the transverse direction and driven by gradients in the wall shear stress. For a deeper insight into the origin of the cross-stream motions and their connection to the turbulence phenomenon, detailed measurements of the turbulent stresses were still needed.

Associated with the secondary flows in non-circular ducts is a mean streamwise component of vorticity; hence by examining the vorticity transport equation, an insight into the origin of these motions can be obtained. For fully developed flow in a straight duct, the mean streamwise vorticity equation reads:

$$\underbrace{\bar{v}\frac{\partial \bar{\Omega}_x}{\partial y} + \bar{w}\frac{\partial \bar{\Omega}_x}{\partial z}}_{C} - \underbrace{\nu\left(\frac{\partial^2}{\partial y^2} + \frac{\partial^2}{\partial z^2}\right)\bar{\Omega}_x}_{D} + \underbrace{\left(\frac{\partial^2}{\partial y^2} - \frac{\partial^2}{\partial z^2}\right)\overline{v'w'}}_{P_1} + \underbrace{\frac{\partial^2}{\partial y\partial z}(\overline{w'^2} - \overline{v'^2})}_{P_2} = 0,$$

(2.7)

where $\bar{\Omega}_x = \frac{\partial \bar{w}}{\partial y} - \frac{\partial \bar{v}}{\partial z}$ is the mean streamwise vorticity, and the prime symbol and overbars represent fluctuating velocity components and time averaging, respectively. y and z represent the wall-normal directions (see Fig. 2.4) while v and w are the corresponding instantaneous velocity components.

The first two terms, C, on the left hand side of Eq. (2.7) represent the convection of streamwise vorticity by the secondary motion itself. Together with the viscous diffusion term, D, these quantities are mainly involved in the redistribution of vorticity within the duct. P_1 and P_2 represent the contribution of the Reynolds cross-stream shear stress and the anisotropy of the cross-stream normal stresses, respectively. They act to either produce or destroy streamwise vorticity. Brundrett and Baines [13], using a hot wire probe, measured the six components of the Reynolds stress tensor

Fig. 2.4 Square geometry and co-ordinate system. x points into the page

as well as all three velocity components, thus obtaining a more detailed description of the turbulence field. From the data collected, an estimate of the terms in the mean streamwise vorticity transport equation was made. Gradients of the Reynolds normal stresses (i.e. P_2) in the ducts cross-section was found to be mainly responsible for the production of streamwise vorticity away from the corners, while the contribution of the Reynolds shear stress term (i.e. P_1) was negligible. On the other hand, viscous diffusion of vorticity occurred nearer to the walls, with the secondary motion acting to transport vorticity from regions of production to regions of diffusion.

While it had been established that the mean secondary flow was at most only a few percent of the primary, there was no consensus on the appropriate scaling parameter. Gessner and Jones [14] observed that when normalised by the bulk or mean centre line velocity, the magnitude of the cross-stream velocities decreased with increasing Reynolds number. To resolve this issue, Launder and Ying [15] conducted experiments in both smooth and rough-walled ducts. The contours of mean axial velocity normalized by the bulk velocity were found to exhibit more bulging towards the corner in the latter than in the former, indicating the presence of larger secondary flows in rough ducts. However, when normalized with friction velocity, the secondary flow profiles in both ducts were found to be the same, thus indicating that the friction velocity is the appropriate normalisation velocity scale.

The aforementioned studies on fully turbulent square duct flow were conducted using hot-wires and pitot tubes. To the best of the author's knowledge, the first non-intrusive measurements in a rectangular duct were conducted by Melling and Whitelaw [16], using LDV. All three mean velocity components as well as five components of the Reynolds stress tensor were obtained. Their results agreed qualitatively with those of previous hot wire studies in terms of the secondary flow pattern and distortions in the contours of mean axial velocity and axial turbulence intensities. Another experimental study on square duct turbulent flow using LDV is that of Escudier and Smith [17], in which measurements were carried out in both Newtonian and non-Newtonian fluids and contours of secondary flow streamlines, mean axial velocity, turbulence intensities and Reynolds shear stress were presented. The skin friction coefficient in Newtonian square duct flow was found to agree with the correlation of Blasius in the turbulent regime. Similarly, the profiles of mean streamwise velocity at various edge locations at $Re \approx 19100$ followed the classical log law: $\bar{u}^+ = 2.5 \, ln(y^+) + 5$ in the near wall region (where the superscript "+" refers normalisation by wall units obtained in terms of the wall shear stress averaged over the perimeter of the duct), thus indicating the validity of the law of the wall irrespective of geometry.

The observed primary and secondary flow fields in the fully turbulent flow of a Newtonian fluid through a square duct are shown in Figs. 2.5 and 2.6 respectively, while Table 2.1 gives a summary of the experimental findings.

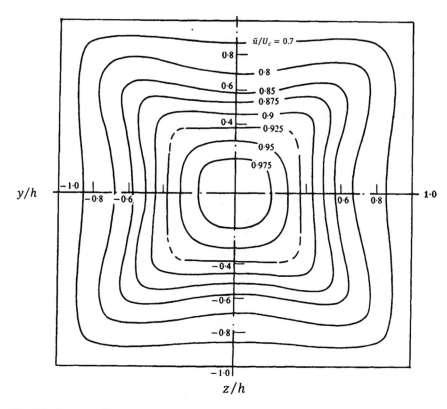

Fig. 2.5 Contours of mean streamwise velocity in a square duct at $Re = 21000$ normalised by the centreline velocity. *Source* Melling and Whitelaw [16]. Reproduced with permission

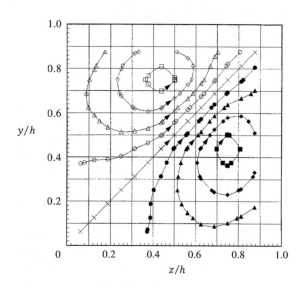

Fig. 2.6 Streamlines of secondary flow in the top right quadrant of a square duct at $Re \approx 19100$: open symbols represent negative values; closed symbols represent positive values. Stream functions, normalised by $U_c D$ (where U_c is the mean axial centreline velocity and D is the duct's diameter), range from -1.6×10^{-3} to 1.6×10^{-3}, with an increment of 4×10^{-4}. Arrows indicate the direction of the secondary flow. Republished with permission from Escudier and Smith [17]; permission conveyed through Copyright Clearance Centre, Inc

Table 2.1 Experiments on turbulent flow of a Newtonian fluid in a square duct

	Author	Fluid	Velocity field data	Instrumentation	Reynolds number (Re)
1	Hoagland [10]	Air	$\bar{u}, \bar{v}, \bar{w}$	Hot wire anemometer and pitot tube	7500–40000
2	Brundrett and Baines [13]	Air	$\bar{u}, \bar{v}, \bar{w}, \overline{u'^2}, \overline{v'^2}, \overline{w'^2},$ $\overline{u'v'}, \overline{v'w'}, \overline{u'w'}$	Hot wire anemometer	41500
3	Gessner and Jones [14]	Air	$\bar{u}, \bar{w}, \overline{v'w'}$	Hot wire anemometer and pitot tube	37500– 150000
4	Launder and Ying [15]	Air	$\bar{u}, \bar{w}$	Hot wire anemometer	34500 and 105000
5	Melling and Whitelaw [16]	Water	$\bar{u}, \bar{v}, \bar{w}, \overline{u'^2}, \overline{v'^2}, \overline{w'^2}$ $\overline{u'v'}, \overline{u'w'}$	Laser Doppler velocimeter	21000
6	Escudier and Smith [17]	Water	$\bar{u}, \bar{v}, \bar{w}, \overline{u'^2}, \overline{v'^2}, \overline{w'^2},$ $\overline{u'v'}$	Laser Doppler velocimeter	19107

2.2.2   Numerical Simulation of Fully Turbulent Square Duct Flows

While new measurements techniques were being developed for the study of complex flows in ducts, progress was also being made on their mathematical modelling. The theoretical treatment of these flows was based on the Reynolds averaged Navier-Stokes equations (RANS). In the RANS approach, each instantaneous flow variable is decomposed into a mean and fluctuating component and averaged over time. This leads to an equation, which in tensor notation reads:

$$\frac{\partial \bar{u}_i}{\partial t} + \bar{u}_j \frac{\partial \bar{u}_i}{\partial x_j} = -\frac{1}{\rho} \frac{\partial \bar{p}}{\partial x_i} + \nu \frac{\partial^2 \bar{u}_i}{\partial x_j^2} - \frac{\partial \overline{u'_i u'_j}}{\partial x_j}, \tag{2.8}$$

where u_i, p and t denote the velocity field, static pressure and time respectively. Equation 2.8 is very similar to the original Navier-Stokes equation except for the additional terms, $\overline{u'_i u'_j}$, the Reynolds stresses, that have been introduced, leading to more equations than unknowns- the closure problem [18].

A classical closure approach models the Reynolds stress tensor using the Boussinesq approximation, i.e. as the product of an eddy viscosity and the mean velocity gradient (see e.g. [19]). It soon became obvious that this class of eddy viscosity models did not contain any mechanism for the generation of secondary flows (see [20]). The anisotropy in the Reynolds stresses play a big role in the generation of these cross-stream motions, hence, second order closure models based on the Reynolds stress transport equation are required. For a detailed overview of modelling work on turbulent flows in non-circular ducts, the reader is referred to the review by Demuren and Rodi [21]. More recent studies using RANS include those by Nisizima [22], Reif and Andersson [23] and Gnanga et al. [24]. These RANS computations yielded results which only agreed qualitatively with experiments, the turbulence properties being either under-predicted or over estimated.

To obtain a highly accurate description of a turbulent flow field, the full Navier-Stokes equation has to be solved directly without incorporating any models. This approach, referred to as direct numerical simulation requires that the computational grids and time steps be fine enough to resolve all spatial and temporal turbulence scales. The required number of grid points scales with Reynolds number to the nine-fourth's power [25], hence at the large Reynolds numbers typical of engineering applications, the computational power required is huge. Therefore, only the simplest flow geometries at relatively low Reynolds numbers have so far been resolved [26].

One of the earliest DNS of square duct turbulent flow was conducted by Gavrilakis [27] at $Re = 2205$ ($Re_\tau = 150$, where $Re_\tau = u_\tau h / \nu$). Using a computational domain long enough to allow for the decorrelation of turbulent statistics and by imposing a periodic boundary condition in the streamwise direction, a fully developed flow was ensured. His results showed good qualitative agreement with experimental findings with regards to bulging of the contour of mean streamwise velocity towards the duct corners and an eight-vortex secondary flow pattern (Fig. 2.7). The skin friction coefficient obtained from the simulation fitted well into the correlation of Jones [28]: $1/\sqrt{f} = 2 \log_{10}(2.25 Re \sqrt{f}) - 0.8$. However, the wall shear stress was found to be non-uniformly distributed across the duct due to the secondary flow field. Specifically, three peaks were observed on each wall; one at the midpoint and the other two close to the corners (see Fig. 2.8). Another key finding from the simulation, which is a consequence of the wall shear stress gradients, was that the streamwise velocity profile along the wall bisector exhibited an overshoot from the classical log-law in the logarithmic layer when normalised by the average (across the duct perimeter) rather than local friction velocity. This is at variance with the experimental findings of Escudier ans Smith [17] which showed the profile for water to follow the log-law and can be potentially attributed to the relatively low Reynolds number of the DNS. Other flow statistics available from the simulation include second and third moments of velocity as well as values of the Reynold shear stress and terms in the equation of mean streamwise vorticity. The anisotropy term, $\frac{\partial^2}{\partial y \partial z}(\overline{w'^2} - \overline{v'^2})$, was found to be the major contributor to the production of vorticity and, contrary to the findings of Brundrett and Baines [13], the contribution of the convection term to vorticity transport was found to be small.

Concurrently with Gavrilakis [27], Huser and Biringen [29] carried out a DNS of turbulent square duct flow at Re_τ of 300. In addition to computing all terms in the equation of mean streamwise vorticity, they also presented the momentum budgets. The results obtained agreed qualitatively with previous experiments and DNS; the observed discrepancies were attributed to differences in Reynolds number. From a quadrant analysis of the Reynolds shear stresses, $\overline{u'v'}$, $\overline{u'w'}$ and $\overline{v'w'}$, it was concluded that the secondary flow in a square duct is induced by the interaction between ejections from two intersecting walls. This is an indication of the role of

Fig. 2.7 Mean streamwise
velocity contours and
secondary velocity vectors.
Source Gavrilakis [27].
Reproduced with permission

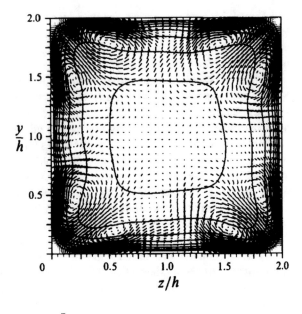

Fig. 2.8 Wall shear stress
distribution: —, standard
run; ...... lower resolution
run. *Source* Gavrilakis [27].
Reproduced with permission

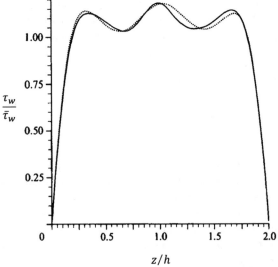

near wall coherent structures in their generation. Joung et al. [30] later showed that
sweeping events dominated in regions where the wall shear stress was maximum
while ejection events dominated in regions of minimum shear stress.

From the foregoing, it is clear that the flow structure in turbulent duct flows shows
a dependence on the Reynolds number. Pinelli et al. [31] investigated this Reynolds
number dependence, focusing on the role of coherent structures in secondary flow
generation and characterization at Re_τ ranging from about 80 to 225. The wall shear

stress distribution was found to be correlated with the number and positioning of the buffer layer streaks. High speed streaks were observed to be positioned in regions of local maxima in the wall shear stress, the first one always being preferentially located close the corners. On the other hand, low-speed streaks were associated with local minima in the wall shear stress distribution. The presence of sidewalls placed a limit on the number of these structures that can be hosted within the duct's span, since the average separation distance between velocity streaks of the same sign in the near-wall region is about 100 in wall units [32]. At Reynolds numbers close to transition (so called "marginally turbulent" regime), the secondary flow pattern was observed to be a direct consequence of the preferential positioning of the coherent structures in the flow, namely: quasi-streamwise vortices and their associated streaks. Studies on marginally turbulent flow will be discussed in detail in Sect. 2.2.3. At higher Reynolds numbers, the number of high and low speed streaks increased, the structures having an equal likelihood of occurring at any given edge location away from the corners, hence the wall shear stress was observed to be more uniformly distributed. A stretching of the secondary flow streamlines towards the corners was also observed.

In recent times, there has been a focus on increasing the Reynolds number attained by DNS. Zhang et al. [33] conducted simulations at shear Reynolds numbers ranging from 150 to 600. The streamwise turbulence intensity was observed to scale in wall units for $Re_\tau > 300$. The duct's centreline velocity, normalized by the bulk velocity was also observed to decrease with increasing Reynolds number due to a reduction in the thickness of the viscous sublayer, resulting in a more uniform velocity profile. Pirozzoli et al. [34], using a compressible flow code, with the turbulent Mach number set to be less than 0.01 in order to closely approximate an incompressible turbulent flow, extended the Reynolds number envelope to Re_τ of 1055 (the highest so far). Contrary to the experimental observations of Gessner and Jones [14], they observed that the intensity of the secondary flow vortices when scaled in outer units was unaffected by variations in Reynolds number. It should be noted that the range of Reynolds numbers examined in the simulations ($Re = 2205$ to 20000) is still well below those considered in the experiment ($Re = 37500$ to 150000). The skin friction coefficient was, however, observed to be in good agreement with the correlation of Jones [28] as well as the Karman-Prandtl theoretical friction law and pipe flow data, thus lending credence to the validity of the hydraulic diameter concept for non-circular ducts.

Owing to the aforementioned difficulties associated with DNS, a more practical computational approach—the large eddy simulation (LES) has been successfully employed in the literature. In this method, only the turbulence structures larger than the grid are directly solved, while the sub-grid scales are modelled. One of the early LES of turbulent flow in a square duct was carried out by Madabhushi and Vanka [35] using a mixed spectral-finite difference scheme for the solution of the Navier-Stokes equation in the grid scale and a Smagorinsky eddy viscosity model for modelling the sub-grid scales. Results of mean axial velocity, secondary flow, mean streamwise vorticity and turbulence kinetic energy at Re_τ of 180 were found to agree qualitatively with the experimental data of Hoagland [10] and Brundrett and Baines [13].

Table 2.2 Numerical simulations of fully turbulent flow of Newtonian fluids in a square duct

S/N	Author	Simulation type	Reynolds number (Re)	Shear Reynolds number (Re_τ)	Dimension of duct	Numerical scheme
1	Gavrilakis [27]	DNS	2205	150	$2h \times 2h \times 20\pi h$	Second order finite difference
2	Huser and Biringen [29]	DNS	5160	300	$2h \times 2h \times 4\pi h$	Mixed spectral-finite difference
3	Joung et al. [30]	DNS	2220	150	$2h \times 2h \times 6\pi h$	Second order finite difference
4	Pinelli et al. [31]	DNS	1077–3500	80–225	$2h \times 2h \times 4\pi h$	Spectral method
5	Zhang et al. [33]	DNS		300–1200	$2h \times 2h \times 4\pi h$	Fourth order spatial descritization scheme
6	Pirozzoli et al. [34]	DNS	2205–20000	150–1055	$2h \times 2h \times 6\pi h$	Fourth order finite difference
7	Madabhushi and Vanka [35]	LES	2905	180	$2h \times 2h \times 4\pi h$	Mixed spectral-finite difference
8	Pattison et al. [36]	LES	2205	150	$2h \times 2h \times 12h$	Lattice Boltzmann method

The observed differences were attributed to the relatively low Reynolds number of the LES compared to those of experiments. In some other studies, the generalised lattice Boltzmann equations, which are based on kinetic theory, rather than the Navier-Stokes equations have been used. One such LES of turbulent square duct flow was conducted by Pattison et al. [36] at $Re_\tau = 150$. The method produced the same secondary flow pattern as the DNS of Gavrilakis [27]. The contour plot of mean axial velocity, was also found to agree with the DNS; however, slight variations were observed along the wall bisector, attributed to differences in grid resolution between DNS and LES. Furthermore, the wall shear-stress was found to follow the same trend observed by Gavrilakis [27], but the values at the corners were significantly lower. Further details on the DNS and LES of turbulent square duct flow are presented in Table 2.2.

2.2.3   Exact Coherent States, Transition to Turbulence and Marginally Turbulent Flows

As for the case of circular pipes, linear stability theory shows that the laminar flow in rectangular channels with aspect ratios less than 3.2 is linearly stable at all Reynolds numbers [37], hence all sufficiently small disturbances should decay. However,

Fig. 2.9 Mean flow field of the travelling wave solution of [40] at $Re = 598.2$. The difference between the mean streamwise velocity and the laminar flow analytical solution is plotted. The data have been normalised by the laminar velocity at the centreline. Contour levels range from -0.570 to 0.099 with an increment of 0.032. Dark colour coding represents negative velocities while light represents positive velocities. Reprinted figure with permission from [40]. Copyright (2009) by the American Physical Society

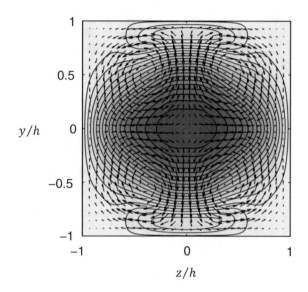

experimental evidence shows an abrupt transition to turbulence, the value of the critical Reynolds number being highly sensitive to initial conditions. A major recent development in the study of laminar-turbulent transition in ducts has been the discovery of alternative solutions to the Navier-Stokes equations. These solutions, which usually take the form of exact coherent structures such as travelling waves were first discovered by Nagata [38] in plane Couette flow and have since been detected also in pipe (see e.g. [39]) and square duct flows (see e.g. [40]). Their occurrence is believed to signal the beginning of turbulence as they are capable of supporting long-lived disordered flow states [41]. Figure 2.9 shows the mean structure of a nonlinear travelling wave solution obtained by [40] for square duct flow at $Re = 598.2$. At different streamwise positions, strong swirling was observed close to either the top or bottom wall, resulting in a streamwise mean flow characterised by four vortices, one in each quadrant, a 90° rotation also being a possible solution. The vortices are, however, oriented in the opposite sense to those observed in the DNS of fully turbulent flow.

Uhlmann et al. [42] obtained a different travelling wave solution at $Re = 471$. The flow structure consisted of counter-rotating vortices (together with their associated low speed streaks) which when averaged in the streamwise direction yielded an eight-vortex secondary flow pattern similar to that observed in the fully turbulent state. By continuing the solution to higher Reynolds number, a skin friction coefficient equal to that in the turbulent state was observed at $Re = 1370$ for the upper branch at a streamwise wave number of 1. The intensity of the secondary flow in the travelling wave solution was also found to be comparable to those in turbulent square duct flow at $Re > 1370$, hence the authors arrived at a conclusion that the computed travelling wave solutions, in some way, played a role in the generation of secondary motions induced by turbulence.

(a)   **(b)**   **(c)**

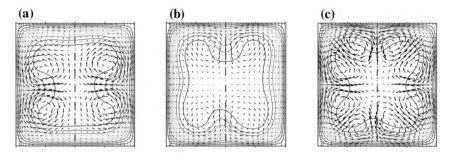

Fig. 2.10 The two flow states of marginally-turbulent flow (**a, b**) and the conventional eight-vortex pattern **c**. Contour lines show the primary mean flow $\bar{u}$ and vectors show the secondary mean flow $\bar{v}, \bar{w}$ for $Re = 1205$: **a** averaging interval $771h/U_b$; **b** a different interval with length $482h/U_b$; **c** long-time integration including both previous intervals ($1639h/U_b$). The wall bisector is indicated by dashed blue lines. Adapted from Fig. 3 of Uhlmann et al. [46]. Reproduced with permission

The lowest Reynolds number so far for which a travelling wave solution has been obtained in a square duct is about 332 [43, 44]. However, numerical simulations reveal that the limiting value of Reynolds number for the onset of self-sustaining turbulence in periodic square-duct flow ranges between $Re = 865$ ($Re_\tau = 65$) and $Re = 1077$ ($Re_\tau = 80$) [45, 46]. More research is therefore needed to determine the role of travelling wave solutions in the transition dynamics.

At Reynolds numbers just above transition, the flow field has been found to behave very differently from what is observed in the fully turbulent regime. The DNS results of Uhlmann et al. [46] are particularly fascinating. So-called marginally turbulent flow was found to be characterized by two states each of which contained a four-vortex secondary flow field rather than the conventional pattern of eight vortices. The vortex pairs, which were associated with a pair of opposite walls, alternated, in time, between two different orientations, the eight-vortex secondary flow pattern being a composition of these two as reproduced in Fig. 2.10. In the marginally-turbulent regime, the duct's width in wall units was just barely large enough to host the minimum number of structures required for self-sustained turbulence, hence at any given time, turbulence activity was concentrated only on two walls while the others remains quiescent. This switching of the flow field between two states had not previously been observed in experiments.

The observed secondary flow pattern was shown to be a direct consequence of buffer layer coherent structures (quasi-streamwise vortices and velocity streaks), whose preferential positioning over the duct walls matched the mean cross-stream flow. It should be noted that this explanation for the origin of secondary flow is only satisfactory in the marginal turbulence regime. As the Reynolds number is increased and more coherent structures are hosted within the duct, the flow dynamics become more complicated.

In this thesis, experimental data on the turbulence field in the "marginal turbulence" regime will be provided in order to validate the DNS results of Uhlmann et al. [46]. The data will be analysed to determine whether a spatial switching effectively exists.

2.3 Wall-Driven Flows in Planar Ducts

Plane Couette flow—the flow between two infinite parallel plates, induced by wall translation- is arguably the simplest type of shear flow. Usually set up either by moving one wall while the other is stationary or by translating both walls with the same speed but in opposite directions, a unique feature of this flow is the existence of a monotonic velocity profile across the entire channel, both in the laminar and turbulent cases. Furthermore, the Navier-Stokes equation for the time-averaged fully developed flow field reduces to the following:

$$\frac{\partial}{\partial y}\left(\mu\frac{\partial \bar{u}}{\partial y} - \rho\overline{u'v'}\right) = 0, \tag{2.9}$$

thus indicating a constant shear stress distribution.

In spite of their apparent simplicity, experiments on wall-driven flows in planar geometries have proven to be very challenging due to the technical difficulties in implementing wall motion in the laboratory without fluid leakage or vibration of moving parts. Numerical simulations of wall-driven flows may be even more computationally expensive to carry out as turbulent plane Couette flow is known to be characterised by very large scale motions (see e.g. [47]), thus necessitating the use of large computational domains. Therefore, there are fewer studies on turbulent plane Couette flow compared to the pressure-driven case [48].

Research questions often addressed in the literature include the stability of the laminar base flow, the dynamics of transition to turbulence and the characteristics of the fully turbulent flow. Like in pressure-driven flow in pipes, plane Couette flow has been shown to be linearly stable at all Reynolds numbers [49, 50] hence transition to turbulence can only be triggered by finite amplitude perturbations. Orszag and Kells [51] showed numerically that a three-dimensional disturbance is required to drive this transition. The onset of chaotic flow is characterised by the appearance of localised turbulent spots or stripes which grow until they occupy the entire flow domain. One of the earliest studies on the growth of turbulent spots in plane Couette flow was conducted by Lundbladh and Johansson [52]. In their direct numerical simulations, initial disturbances of large amplitudes were introduced into the laminar flow and their evolution in time, monitored. A key finding from the study was that localised turbulent spots, mostly elliptical in shape could grow into sustained turbulence only for $Re_w > 375$ ($Re_w = U_w h/\nu$, where U_w is the velocity of a moving wall). Other experimental and numerical studies have since been conducted to further investigate the transition phenomenon (see e.g. [52–54]).

Experimental facilities have been designed to allow for visualisation of the flow field. A configuration having counter-moving walls (with wall motion generated by a transparent endless belt), resulting in a zero mean flow across the channel has mainly been adopted. This arrangement ensures that turbulent spots remain stationary in the laboratory frame of reference for a long time, making them easier to study, as opposed to having to track them as they are advected along the channel. The first flow visualisation experiments on transition were carried out by Tillmark and Alfredsson [53] in a channel of aspect ratio 36:1. They perturbed the laminar base flow by introducing a bubble at the bottom of the channel and monitored the evolution of the disturbance created as the bubble rose to the surface. Figure 2.11 shows the growth of a typical turbulent spot. The critical[1] Reynolds number (Re_w) for the sustenance of a turbulent state was found to be about 360—very close to that obtained from the simulations of Lundbladh and Johansson [52]. Another key finding from the study was the detection of waves travelling away from the turbulent regions. A similar observation was made by Daviaud et al. [54].

Due to the sub-critical nature of the transition, the critical Reynolds number is very sensitive to the amplitude of the disturbance introduced into the flow. Dauchot and Daviaud [55] studied the dependence of the critical perturbation amplitude on the Reynolds number, observing a power law behaviour. From their results, the critical Reynolds number was estimated to be about 325. Shi et al. [56] showed that the localised turbulent structures are transient in nature, increasing in size temporarily before seeding new ones and eventually decaying. The spreading rate of the turbulent patches was observed to increase with Reynolds number, eventually outweighing the decay rate when the turbulence became self-sustaining. The splitting and decay time scales of turbulent spots were observed to follow an exponential distribution, with a Reynolds number of 325 corresponding to the point where there was a balance between both processes. From these results, it can be concluded that the smallest Re_w at which a turbulent state can be maintained in plane Couette flow is about 325.

The characteristics of wall driven flows in planar ducts in the fully turbulent regime are also still being actively researched. The early studies [57–62] provided data on the mean velocity profile, turbulence intensities, Reynolds stress distribution, the skin friction coefficient and other turbulence statistics. The configuration with a stationary wall and an opposite translating wall was mostly adopted. In these studies, the turbulence intensities in the central region of the duct were found to be much higher than those in pressure-driven flow. A non-zero velocity gradient was also observed in the channel mid-plane, resulting in a finite production rate of turbulent kinetic energy there. There are, however, conflicting results on the asymptotic behaviour of this gradient as Reynolds number tends to infinity, some studies showing a slow decrease in the slope [63] and others indicating that it becomes constant [64].

Direct numerical simulations of plane Couette flow reveal that the turbulence structure in the core region is very different from that in pressure-driven flow.

---

[1]The same value of critical Reynolds number was obtained when the experiment was initiated from a fully turbulent state and the Reynolds number was gradually reduced until turbulence could no longer be sustained.

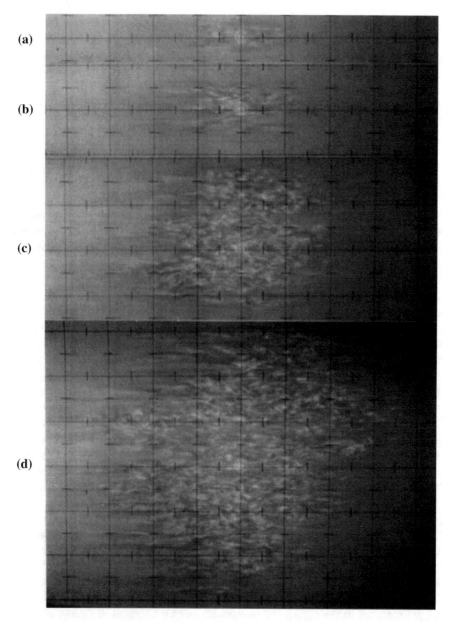

Fig. 2.11 Development of a turbulent spot at $Re_w = 405$. **a, b, c** and **d** show the spot after a time of 25, 95, 210 and 390 h/U_w respectively. Adapted from Fig. 6 of Tillmark and Alfredsson [53]. Reproduced with permission

Lee and Kim [65] discovered the existence of large-scale motions in the form of quasi-streamwise vortices and their associated streaks, extending across the entire channel length. These structures were found to be very persistent in space and time. Although the computational domain employed was not large enough to obtain a decorrelation of the turbulence statistics, the findings have been confirmed by subsequent experimental studies [66, 67] as well as numerical simulations in large computational boxes [47, 48, 68, 69]. The large structures, 4–5 h wide and 40–65 h long [67, 68] have been shown to be very sensitive to an imposed system rotation [47] and to make a substantial contribution to the turbulent shear stress in the channel core [48]; their origin is however not yet fully understood. In the analysis by Papavassiliou and Hanratty [70], the vortical structures were approximated as secondary flows, and shown to be sustained by energy from small scales (reverse energy cascade). This finding is contrary to the well-known energy cascade from large to small scales and has not been confirmed by other studies.

Due to the huge computational resources required for the simulations earlier Couette flow DNS were conducted at shear Reynolds numbers less than 170. Experiments by Kitoh et al. [66] revealed that the turbulent statistics in this regime is affected by low Reynolds number effects. For $Re_\tau < 150$, a decrease in the value of the additive constant, B, in the classical log-law was observed. The peak value of the streamwise turbulence intensity was also found to drop sharply. As a result of the three-dimensional effect introduced by the rolls, the skin friction coefficient was observed to vary by up to 20% in the spanwise direction. More recent numerical studies have focused on investigating the flow at higher Reynolds numbers in order to obtain an insight into the asymptotic behaviour of the turbulence field. Avsarkisov et al. [69] conducted direct numerical simulations at shear Reynolds numbers ranging from 125 to 550, presenting data on the turbulence statistics and also providing a visualisation of the coherent structures. Large scale motions observed in previous studies were shown to be present at the Reynolds numbers investigated. The longest structures were found to be organised as counter-rotating vortices. Plots of streamwise turbulence intensity (u_{rms}^+) showed a lack of collapse with increasing Reynolds number. The non-dimensional velocity gradient, ψ at the channel centreline (referred to as the slope parameter) was also observed to decrease with Reynolds number but there was not enough data to estimate the behaviour in the limit as $Re_\tau \to \infty$.

The DNS results of Pirozzoli et al. [48] at Re_τ ranging from 171 to 968 show similar trends in the flow statistics as observed by Avsarkisov et al. [69]. However, the streamwise velocity variance ($\overline{u'^2}$) at $Re_\tau = 986$ was found to exhibit an outer peak[2] at $y/h \approx 0.33$ (corresponding to $y^+ \approx 300$). At this location, an excess in the production of turbulent kinetic energy over the dissipation rate was observed. This excess energy (about 20% of the one at the inner peak location) was found to be redistributed to the rest of the channel mainly by turbulent diffusion. Another major finding from the simulations of Pirozzoli et al. [48] was the absence of a genuine logarithmic near-wall layer in plane Couette flow. Although the streamwise velocity profiles in wall units appeared to follow the classical log law with $\kappa = 0.41$ and $B = 5$, plots of the log-indicator function $y^+ \frac{d\bar{u}^+}{dy^+}$ against non-dimensional duct

---

[2]It should be noted that the presence of an outer peak in the distribution of $\overline{u'^2}$ in wall-bounded turbulent flows had only been shown by experiments on pressure driven flow at $Re_\tau > 20000$ [71].

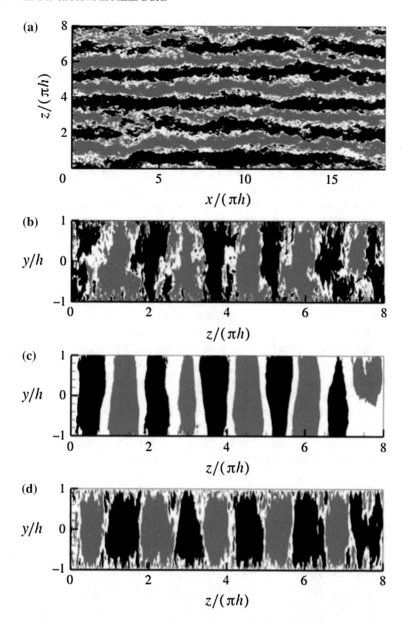

Fig. 2.12 Large scale motions in turbulent plane Couette flow: **a** and **b** are the contours of instantaneous streamwise velocity at the centre plane and a cross-sectional plane respectively; **c** and **d** show the streamwise averages of the streamwise velocity and vorticity respectively. Negative values are shown in black and positive in grey. *Source* Pirozzoli et al. [48]. Reproduced with permission

height, y^+, did not exhibit any plateau region. Visualisations of the instantaneous velocity fields showed, clearly, the organisation of the flow into coherent high and low speed streaks, whose streamwise extents were comparable to the channel length. These eddies were observed to occupy the entire channel height, extending from one wall to the other (see Fig. 2.12)

In the studies discussed so far, the effect of side walls on the structure of the Couette flow was not considered. It will be interesting to investigate the influence of secondary flows generated at the corners of a duct on the large scale structures induced by wall translation. To the best of the author's knowledge, the only studies where secondary motions in wall-driven flow have been examined are those by Lo and Lin [73] and Hsu et al. [72] on turbulent Couette-Poiseuille flow in a square duct with a single moving wall. In these large eddy simulations the flow field and turbulence characteristics close to the stationary wall were observed to be very similar to those of Poiseuille flow with marked differences occurring as the moving wall is approached. The ratios of wall to bulk velocity (r), considered ranged between 0 (Poiseuille flow) to 3.15 (purely wall driven flow). For low and high values of r, a logarithmic region was found to exist in the velocity distribution close to the moving wall, while departures from the log-law occurred at intermediate values of r ($r = 0.91$–1.6). On the other hand, the velocity profiles at the stationary wall followed the log-law in all the cases considered. As a result of the modification of the strain rate by the moving wall, the maximum production of turbulent kinetic energy was different from that at the stationary wall. The secondary flow pattern was observed to be significantly influenced by the presence of a translating wall. As the wall velocity increased, a merger of the secondary vortices occurred and the cross-stream flow changed from an eight-vortex pattern to one characterised by six vortices, with symmetry only about the moving-wall bisector (see Fig. 2.13). Examination of the terms in the equation of mean streamwise vorticity, revealed the importance of the gradients of Reynolds normal and shear stresses in the production of the vortices.

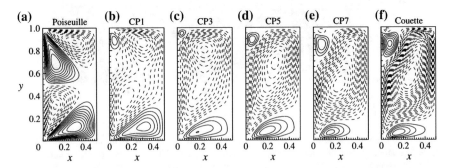

Fig. 2.13 Secondary flow streamlines in square duct Couette-Poiseuille flow at different ratios of wall to bulk velocity (r). CP1, CP3, CP5 and CP7 show the secondary flow at $r = 0.6$, 1.14, 1.37 and 2.28 respectively. Flow is symmetric about the wall bisector, $z = 0.5$, hence only the left half of the domain is shown. The top wall is translated in the direction of the bulk flow while the other walls are kept stationary. *Source* Hsu et al. [72]. Reproduced with permission

More research is needed to improve our understanding of wall-driven flows in ducts where side-wall effects cannot be neglected. Specifically, the flow field needs to be investigated over a wide range of Reynolds numbers spanning from the "marginally-turbulent" regime to the fully-turbulent state. The effect of side walls on the transition to turbulence also needs to be determined. In this thesis, an attempt is made to address these issues. DNS of Couette flow in a square duct is conducted to determine the critical Reynolds number for transition to turbulence and also to characterise the turbulence field at relatively low Reynolds numbers. In addition, a new facility is designed and constructed to facilitate the experimental investigation of Couette-Poiseuille flows.

2.4 Duct Flows with Polymer Additives

It is well known that long-chain flexible polymers having high molecular weight possess excellent drag-reduction (DR) capabilities in turbulent flow when added to a Newtonian solvent even at minute concentrations. This phenomenon was first discovered by Toms [74] while observing the flow of monochlorobenzene mixed with a polymer- polymethylmethacrylate. The resulting non-Newtonian solution gave rise to a lower pressure drop compared to the flow of the pure solvent. Other types of additives, such as surfactants, rigid fibres and bubbles have since been found to produce similar effects. Potential benefits range from the decrease in cost of oil transport through pipelines and increased efficiency of fire fighting equipment as well as irrigation and drainage systems, to the reduction in skin friction drag over ships and submarines; hence a large number of studies have been conducted in order to obtain an insight into the underlying mechanism, with researchers approaching the problem from theoretical, experimental and numerical perspectives. However, seven decades after its discovery, a full understanding of the mechanism for the drag reduction phenomenon is yet to be obtained. This is not surprising, as the problem requires a detailed grasp of the dynamics of turbulent flows and rheology of polymeric fluids—two complex subject areas. Periodic overviews of the literature published on the subject can be found in the reviews by Lumley [75], Virk [76], Den Toonder et al. [77], White and Mungal [78] and Procaccia et al. [79] among others. In this section, a summary of the important findings on the phenomenon will be presented, with a focus on the turbulent flow of polymer solutions in ducts, but first, a brief introduction to the rheology of dilute and semi-dilute polymer solutions is given.

2.4.1 Rheology of Dilute and Semi-dilute Polymer Solutions

A polymer solution can be regarded as dilute if the individual molecules are so far apart that they do not interact [80]. As the polymer concentration, c, is increased, the spacing between the molecules decrease and eventually, when a certain value, referred to as the critical overlap concentration, c^* is exceeded, the polymer chains become entangled. The solution is said to be semi-dilute in the regime where $c \sim c*$. For values of $c \gg c^*$, the solution is regarded as being concentrated. An estimate of c^* for a polymer solution can be obtained from the correlation of Graessley [81]:

$$c^* = \frac{0.77}{[\eta]}, \qquad (2.10)$$

where the intrinsic viscosity, $[\eta]$ is given by

$$[\eta] = \lim_{c \to 0} \frac{\eta_{sp}}{c}. \qquad (2.11)$$

In 2.11, the specific viscosity, η_{sp}, gives a measure of the increase in the shear viscosity of a solution relative to the solvent as a result of the addition of polymer molecules:

$$\eta_{sp} = \frac{\mu_{solution} - \mu_{solvent}}{\mu_{solvent}}. \qquad (2.12)$$

The viscosities in Eq. 2.12 are those in the zero shear rate limit.

In drag reduction applications, only dilute and semi-dilute solutions are of interest. The polymer molecules, when added to a Newtonian solvent, change the fluid's rheology, thus making the resulting solution non-Newtonian. A non-Newtonian fluid can be defined as one which shows a deviation from any of the Newtonian behaviour listed below (see [82]):

(a) Shear viscosity is constant.
(b) In a simple shear flow, the only stress generated is τ_{xy}.
(c) The stress in the fluid vanishes immediately upon cessation of shearing.
(d) Viscosity measured in uniaxial elongational flow is always thrice[3] the value in simple shear flow.

Examples of non-Newtonian behaviour can be found in Fig. 2.14. Most polymer solutions studied in the literature are shear-thinning (i.e. their viscosity decreases with increasing shear rate). The power-law model (Eq. 2.13) is arguably the simplest constitutive law which predicts this behaviour:

$$\tau_{xy} = K\dot{\gamma}^n, \qquad (2.13)$$

where K, with units of Pa.s^n is referred to as the "consistency", n is the power-law index and $\dot{\gamma}$ is the shear rate. From Eq. 2.13, an expression for the "apparent viscosity" (viscosity assuming a linear relationship between stress and strain) can be derived:

$$\mu = K\dot{\gamma}^{n-1}, \qquad (2.14)$$

---

[3]The Trouton ratio, Tr is the ratio of elongational to shear viscosity. For a non-Newtonian fluid, $Tr > 3$.

Fig. 2.14 Variation of shear stress with shear rate in Newtonian and non-Newtonian fluids. (i), Bingham plastic; (ii), shear-thinning (pseudoplastic); (iii), Newtonian; (iv), shear-thickening (dilatant). Adapted from: Erhard et al. [83]

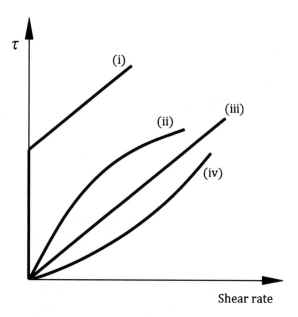

where $n = 1$ represents a Newtonian fluid, and $n > 1$ and $n < 1$ indicate shear thickening and shear thinning respectively. A draw back of the power-law model is that it cannot capture fluid behaviour in the limits of zero or infinite shear rates; hence other constitutive relations have been developed. One such equation is the Carreau-Yasuda model [84, 85]

$$\frac{\mu - \mu_\infty}{\mu_0 - \mu_\infty} = \frac{1}{[1 + (K_{cy}\dot{\gamma})^m]^{n/m}}, \tag{2.15}$$

where μ_∞ represents the viscosity at infinite shear rate, μ_o is the zero shear rate viscosity, K_{cy} is a constant with a dimension of time, representing the onset of shear-thinning, m is a parameter introduced by Yasuda et al. [85] and n, a constant indicating the degree of shear thinning.

Polymer solutions also exhibit elastic properties, indicated by departure from rules (b) to (d) above. When subjected to an elongational deformation, their extensional viscosity increases greatly [86]. Two non-dimensionless numbers that can be used to characterise viscoelastic effects are the Deborah (De) and Weissenberg (Wi) numbers. The Deborah number is defined as the ratio of the relaxation time,[4] λ to the characteristic time scale, T_o of a process. It is a dimensionless measure of the rate at which flow conditions change, hence in flows such as steady simple shear, where the time scale for the deformation is infinite, $De = 0$ [87].

---

[4]The relaxation time, λ, is the time required for a stretched polymer to return to an equilibrium state.

The Weissenberg number, Wi is the ratio of elastic to viscous forces. This is equivalent to a product of the relaxation time and a characteristic deformation rate ($\dot{\gamma}$). A prefactor can also be included depending on the constitutive model used. In a simple shear flow, for example, an estimate of the elastic force can be obtained from the first normal-stress difference ($N_1 = \sigma_{xx} - \sigma_{yy}$) as $N_1 = 2\lambda\mu\dot{\gamma}^2$, if the upper-convected Maxwell constitutive equation is used, while the viscous force can be obtained from the shear stress, $\tau_{xy} = \mu\dot{\gamma}$. The Weissenberg number is thus given by [87]

$$Wi = \frac{2\lambda\mu\dot{\gamma}^2}{\mu\dot{\gamma}} = 2\lambda\dot{\gamma}. \qquad (2.16)$$

2.4.2 Findings on Polymer Drag Reduction from Experiments and Numerical Simulations

Experimental studies on turbulent drag reduction have provided information mainly on the gross flow properties and turbulence statistics. To eliminate errors due to flow disturbances introduced by measurement probes (see e.g. [88]), most of the velocity field measurements reported in the literature have been obtained using LDV and to a lesser extent, PIV—both non-intrusive techniques. Data often presented include profiles of mean streamwise velocity, streamwise and wall-normal velocity fluctuations, and Reynolds stresses. One of the most important findings on polymer drag-reduction in wall-bounded turbulent flows is the existence of a maximum drag-reduction (MDR) asymptote [76]. Data from gross flow studies plotted in Prandtl-Von Karman co-ordinates ($1/\sqrt{f}$ vs. $Re_\delta\sqrt{f}$, where f is the Fanning friction factor and Re_δ, the Reynolds number based on duct hydraulic diameter) shows that at onset of DR, there is a slope increment from the Newtonian Prandtl-Karman law which is proportional to polymer concentration. Eventually, an upper bound—MDR—is reached for sufficiently large values of Re_δ and/or polymer concentration (see Fig. 2.15). This state is a universal feature of drag reduction, independent of solvent type, polymer concentration, or molecular structure of the polymer additive. The mean velocity profile in wall units for drag-reducing flows have also been found to lie between the log law for Newtonian fluids and MDR asymptote [76] given by the expression: $\bar{u}^+ = 11.7\,ln(y^+) - 17$. It should be noted that there is some scatter in the data for flows nominally at MDR (see [89]), hence Virk's asymptote is approximate.

In drag-reducing flows, the wall-normal and spanwise velocity fluctuations have been shown to always decrease. The streamwise velocity fluctuations, on the other hand, are observed to increase (when these velocities are normalised by the friction velocity), and their maximum value is shifted away from the wall with increasing drag reduction. At high polymer concentrations, however, the maximum fluctuations return to values comparable to those of the Newtonian solvent. Analyses of the turbu-

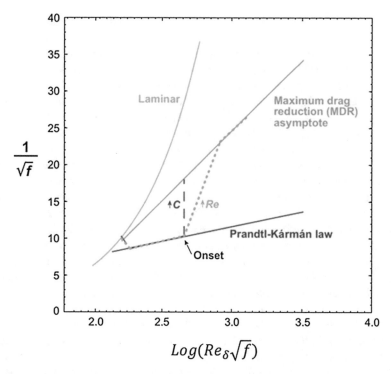

Fig. 2.15 The maximum drag-reduction asymptote. Adapted from Fig. 3 of White and Mungal [78]. Republished with permission; permission conveyed through Copyright Clearance Center, Inc

lent energy spectra also indicate increased damping of high wavenumber fluctuations with increasing drag reduction [90].

Two distinct flow regimes have been identified. For values of DR less than about 40% (usually referred to as the low drag reduction regime), the streamwise velocity in wall units is observed to follow Virk's ultimate profile up to a y^+ after which it becomes roughly parallel to the Newtonian log-law. This logarithmic region, is referred to as the Newtonian plug and it is characterised by a velocity increment. The thickened buffer layer between the Newtonian plug and the viscous sublayer is refered to as the elastic sublayer [76]; interactions between the polymer molecules and the flow in this region is believed to be crucial to the drag reduction phenomenon. Its thickness increases with increasing level of drag reduction. Polymer stretching has been shown to bring about a reduction in momentum flux from the bulk to the wall [79]. The Reynolds shear stress is observed to be reduced below the value in a Newtonian turbulent flow, leading to a stress deficit, accounted for by the generation of polymer stresses [90–92]. The polymer stress is usually estimated indirectly from the total stress using Eq. 2.17),

$$\bar{\tau}_{xy} = \mu \frac{d\bar{u}}{dy} - \overline{\rho u'v'} + \overline{\tau_{xy}^p}, \tag{2.17}$$

where $\bar{\tau}_{xy}$ and $\overline{\tau_{xy}^p}$ represent the mean values of the total and polymer stresses respectively.

For values of drag reduction greater than about 40%, (referred to as the high drag reduction (HDR) regime) the streamwise velocity profiles cease to be parallel to the Newtonian log law, rather they are tilted towards the MDR profile. A re-examination of the logarithmic behaviour of the mean velocity profiles in polymeric flows by White et al. [93] revealed the absence of a true logarithmic region in the HDR regime, although visually fitting a line to the velocity profiles showed otherwise; hence the velocity profiles close to MDR correspond to a thickened buffer layer. The Reynolds shear stress at HDR is observed to be strongly diminished. The data of Warholic et al. [90] show a vanishing of the Reynolds shear stress close to MDR. However, measurements by Ptasinski et al. [91] indicate that although it is greatly reduced, its value remains finite. Therefore, it is very clear that the turbulence sustenance mechanism in the high drag reduction regime is very different from that in a Newtonian flow, with polymer stresses playing a very important role.

A major challenge faced by experimental studies, is the mechanical degradation of polymer molecules, leading to poor repeatability and large uncertainties in data especially in the region between onset and MDR. In recent times, numerical simulations of viscoelastic turbulent flows have been conducted using simple models, such as the finite-extensibility non-linear elastic dumbbell model with the Peterlin approximation (FENE-P), in which a polymer molecule is treated as a pair of spherical beads connected by an elastic spring [94–96]. The polymer stress, given by Eq. 2.19, is incorporated into the momentum equation (2.18) and the equation for the evolution of the conformation tensor, c_{ij} (Eq. 2.20) is also solved (see [78]).

$$\rho \frac{\partial u_i}{\partial t} + \rho u_j \frac{\partial u_i}{\partial x_j} = -\frac{\partial p}{\partial x_i} + \mu \frac{\partial^2 u_i}{\partial x_j^2} + \frac{\partial \tau_{ij}^p}{\partial x_j} \tag{2.18}$$

$$\tau_{ij}^p = \frac{1}{Wi} \left(\frac{c_{ij}}{1 - c_{kk}/\mathcal{L}^2} - \delta_{ij} \right) \tag{2.19}$$

$$\frac{\partial c_{ij}}{\partial t} + u_k \frac{\partial c_{ij}}{\partial x_k} = c_{kj} \frac{\partial u_i}{\partial x_k} + c_{ik} \frac{\partial u_j}{\partial x_k} - \tau_{ij}^p \tag{2.20}$$

In the above expressions, $Wi = \lambda \dot{\gamma} = \lambda u_\tau^2 / \nu$ is the Weissenberg number based on the shear rate, $\dot{\gamma}$, at the wall, $\mathcal{L}$ is the maximum extension of the strings and δ_{ij} is the Kronecker delta. So far the FENE-P model has reproduced some universal features of polymer drag-reduction such as the MDR asymptote (see [89]), but it is also known to be unable to correctly predict pressure-losses in laminar contraction flows for example [97] and overestimate viscoelastic stresses in turbulent flow [98].

In addition to investigating the effect of polymer molecules on the turbulence statistics, their interaction with near-wall coherent structures have also been studied using the FENE-P model. The addition of polymer molecules to a Newtonian fluid has been observed to bring about a decrease in streamwise vorticity fluctuations [94] and to surpress near-wall vortices [99, 100]. These findings are consistent with the experimental observation of Escudier and Smith [17], on the weakening of secondary flows in a square duct, and the recent square duct polymer flow DNS of Shahmardi et al. [101] which shows a decrease in the magnitude of the wall-normal velocity component. Analysis of the turbulent kinetic energy budgets show that the pressure strain, responsible for the redistribution of energy from the streamwise to the spanwise and wall-normal directions, is much lower than in a Newtonian flow [102]. This explains the increase in the streamwise velocity fluctuations. Polymer stress work was observed to transfer energy from the turbulence to the polymer molecules upon stretching. However, there are conflicting results as to whether this energy is released back to the flow or dissipated when the polymer relaxes (see [79, 102, 103]). This is possibly due to the fact that the FENE-P model does not capture coil-stretch hysteresis in polymer chains.

2.4.3 Theories and Prediction

One conclusion on which there is a consensus among researchers is that the drag-reduction phenomenon is as a result of the dynamical interaction between polymer molecules and turbulence. This interaction begins when their relaxation time becomes comparable to a characteristic time scale of the flow,[5] leading to considerable stretching of the polymer molecules. A detailed analysis by Lumley [105] shows that this onset criterion is given by the so-called "coil-stretch" transition, and in a purely-extensional flow at strain rate $\dot{\varepsilon}$, this occurs for $Wi = \lambda\dot{\varepsilon} \geq 1/2$. In a chaotic, turbulent-like flow of mixed shear and extensional flow, Stone and Graham [98] have shown that significant stretching occurs when the product of the relaxation time and the largest Lyapunov exponent approaches $1/2$.

Theoretical explanations for the drag reduction phenomenon have been based either on the viscous effects induced by polymer stretching (see [75, 105]) or on elastic effects (see [106]). According to Lumley [105], polymers are effectively stretched in regions of the flow where the strain rates are high and vorticity is low. These conditions typically exist in the buffer layer in wall bounded flows. Polymer stretching in extensional flows leads to a huge increase in effective viscosity [86], causing a thickening of the buffer layer and suppression of turbulent fluctuations, thus bringing about drag reduction. This theory is not without its criticisms. It has been argued that the fluctuating strain rates observed in a turbulent flow preclude the full stretching of the polymers thus leading to only a small increase in viscosity [106]. Other proponents of the theory of drag reduction based on viscous effects include Ryskin [107] and Lvov et al. [108]. The latter showed that an effective viscosity profile increasing linearly with distance from the wall could bring about a greater

---

[5]The size of a typical polymer molecule in the fully extended state is much smaller than the Kolmogorov length state, hence onset of drag reduction cannot be explained by a matching of length scales (see [79, 104]).

reduction in Reynolds stress compared to the increase in viscous drag, thus resulting in drag reduction.

An explanation for drag reduction based on elastic behaviour of polymers was given by Tabor and De Gennes [106], with modifications to the theory introduced by Sreenivasan and White [109]. While the turbulent energy reduces with decreasing length scale, the elastic energy stored in the polymer molecules was shown to increase (as a result of increased stretching). Eventually, the energy stored in the polymers becomes comparable to that of the turbulence, at a certain length scale greater than the Kolmogorov length scale, leading to a truncation of the energy cascade. This termination of the non-linear turbulent energy transport mechanism at a length scale greater than the smallest one is believed to lead to a thickening of the buffer layer, thus reducing drag [109]. This theory has been criticised due to the fact that it was formulated in the context of homogeneous, isotropic turbulence and may not be relevant in wall bounded flows, where the turbulence is far from being isotropic.

Given the important role played by polymer stretching, an understanding of the extensional rheological properties of polymer solutions has long been felt to be crucial to obtaining a better insight into the drag-reduction mechanism. Until the fairly recent introduction of the commercial Capillary Breakup Extensional Rheometer (CaBER) [110, 111], accurate measurements of these properties have been elusive to experimentalists due to the dilute[6] nature of the polymer solutions and the difficulty in creating a truly extensional flow for rheometric measurements. A few studies exist where an attempt has been made to quantitatively relate drag reduction to polymer rheological properties in shear or extension, with little success [113–115]. One approach has been to estimate polymer relaxation time from shear rheometric measurements but such experiments are equally challenging for dilute polymers especially in aqueous solution [116, 117]. Another unsuccessful approach to find a correlation was to directly measure the extensional viscosity using opposed-nozzle devices (e.g. [118]), but the flow fields created by such configurations have been shown to be corrupted by shear and inertia, causing the measurements to differ form the true material property [119]. Thus, quantitative predictions of drag-reduction level from a measurable material property have remained elusive. To resolve this issue, a correlation which relates DR to polymer relaxation time measured using CaBER will be developed in this thesis.

References

1. Bradshaw P, Huang GP (1995) The law of the wall in turbulent flow. Proc R Soc Lond A 451(1941):165–188
2. Klewicki J, Fife P, Wei T (2009) On the logarithmic mean profile. J Fluid Mech 638:73–93
3. Marusic I, Monty JP, Hultmark M, Smits AJ (2013) On the logarithmic region in wall turbulence. J Fluid Mech, 716

---

[6]Drag reduction has been reported for polymer concentrations as low as 0.02 ppm [112].

4. El Khoury GK, Schlatter P, Noorani A, Fischer PF, Brethouwer G, Johansson AV (2013) Direct numerical simulation of turbulent pipe flow at moderately high Reynolds numbers. Flow Turbul Combust 91(3):475–495
5. Johnston J (1975) Internal flows. In: Turbulence. Topics in applied physics, vol 12. Springer, pp 109–169
6. Bradshaw P (1987) Turbulent secondary flows. Annu Rev Fluid Mech 19(1):53–74
7. Nikuradse J (1926) Untersuchungen über die geschwindigkeitsverteilung in turbulenten strö-mungen. VDI Forsch, p 281
8. Prandtl L (1927) Turbulent flow. NACA Tech Memo 435
9. Stanković BD, Belošević SV, Crnomarković N, Stojanović AD, Tomanović ID, Milićević AR (2016) Specific aspects of turbulent flow in rectangular ducts. Therm Sci 00:189
10. Hoagland LC (1960) Fully developed turbulent flow in straight rectangular ducts: secondary flow, its cause and effect on the primary flow. PhD thesis, Massachusetts Institute of Technology, Cambridge
11. Durst F, Melling A, Whitelaw JH (1981) Principles and practice of laser Doppler anemometry. Academic Press, London
12. Adrian RJ, Westerweel J (2011) Particle image velocimetry. Cambridge University Press, New York
13. Brundrett E, Baines WD (1964) The production and diffusion of vorticity in duct flow. J Fluid Mech 19(03):375–394
14. Gessner FB, Jones JB (1965) On some aspects of fully-developed turbulent flow in rectangular channels. J Fluid Mech 23(4):689–713
15. Launder BE, Ying WM (1972) Secondary flows in ducts of square cross-section. J Fluid Mech 54(02):289–295
16. Melling A, Whitelaw JH (1976) Turbulent flow in a rectangular duct. J Fluid Mech 78(02):289–315
17. Escudier M, Smith S (2001) Fully developed turbulent flow of non-Newtonian liquids through a square duct. Proc R Soc A 457(2008):911–936
18. Tennekes H, Lumley JL (1972) A first course in turbulence. MIT press
19. Krajewski B (1970) Determination of turbulent velocity field in a rectilinear duct with non-circular cross-section. Int J Heat Mass Transf 13(12):1819–1823
20. Launder BE, Ying WM (1973) Prediction of flow and heat transfer in ducts of square cross-section. Proc Inst Mech Eng 187(1):455–461
21. Demuren AO, Rodi W (1984) Calculation of turbulence-driven secondary motion in non-circular ducts. J Fluid Mech 140:189–222
22. Nisizima S (1990) A numerical study of turbulent square-duct flow using an anisotropic k-model. Theor Comput Fluid Dyn 2(1):61–71
23. Reif BAP, Andersson HI (2002) Prediction of turbulence-generated secondary mean flow in a square duct. Flow Turbul Combust 68(1):41
24. Gnanga H, Naji H, Mompean G (2009) Computation of a three-dimensional turbulent flow in a square duct using a cubic eddy-viscosity model. C R Mec 337(1):15–23
25. Coleman GN, Sandberg RD (2010) A primer on direct numerical simulation of turbulence-methods, procedures and guidelines
26. Jiménez J, Kawahara G (2012) Dynamics of wall-bounded turbulence. Cambridge University Press, pp 221–268
27. Gavrilakis S (1992) Numerical simulation of low-Reynolds-number turbulent flow through a straight square duct. J Fluid Mech 244:101–129
28. Jones OC (1976) An improvement in the calculation of turbulent friction in rectangular ducts. Trans ASME J: J Fluids Eng 98(2):173–180
29. Huser A, Biringen S (1993) Direct numerical simulation of turbulent flow in a square duct. J Fluid Mech 257:65–95
30. Joung Y, Choi S, Choi J (2007) Direct numerical simulation of turbulent flow in a square duct: analysis of secondary flows. J Eng Mech 133(2):213–221

31. Pinelli A, Uhlmann M, Sekimoto A, Kawahara G (2010) Reynolds number dependence of mean flow structure in square duct turbulence. J Fluid Mech 644:107–122
32. Kim J, Moin P, Moser R (1987) Turbulence statistics in fully developed channel flow at low Reynolds number. J Fluid Mech 177:133–166
33. Zhang H, Trias FX, Gorobets A, Tan Y, Oliva A (2015) Direct numerical simulation of a fully developed turbulent square duct flow up to Reτ= 1200. Int J Heat Fluid Flow 54:258–267
34. Pirozzoli S, Modesti D, Orlandi P, Grasso F (2018) Turbulence and secondary motions in square duct flow. J Fluid Mech 840:631–655
35. Madabhushi RK, Vanka SP (1991) Large eddy simulation of turbulencedriven secondary flow in a square duct. Phys Fluids A 3(11):2734–2745
36. Pattison MJ, Premnath KN, Banerjee S (2009) Computation of turbulent flow and secondary motions in a square duct using a forced generalized lattice boltzmann equation. Phys Rev E 79(2):026704
37. Tatsumi T, Yoshimura T (1990) Stability of the laminar flow in a rectangular duct. J Fluid Mech 212:437–449
38. Nagata M (1990) Three-dimensional finite-amplitude solutions in plane Couette flow: bifurcation from infinity. J Fluid Mech 217:519–527
39. Faisst H, Eckhardt B (2003) Traveling waves in pipe flow. Phys Rev Lett 91(22):224502
40. Wedin H, Bottaro A, Nagata M (2009) Three-dimensional traveling waves in a square duct. Phys Rev E 79:065305
41. Hof B, van Doorne CWH, Westerweel J, Nieuwstadt FTM, Faisst H, Eckhardt B, Wedin H, Kerswell RR, Waleffe F (2004) Experimental observation of nonlinear traveling waves in turbulent pipe flow. Science 305(5690):1594–1598
42. Uhlmann M, Kawahara G, Pinelli A (2010) Traveling-waves consistent with turbulence-driven secondary flow in a square duct. Phys Fluids 22(8):084102
43. Okino S, Nagata M, Wedin H, Bottaro A (2010) A new nonlinear vortex state in square-duct flow. J Fluid Mech 657:413–429
44. Okino S, Nagata M (2012) Asymmetric travelling waves in a square duct. J Fluid Mech 693:57–68
45. Biau D, Bottaro A (2009) An optimal path to transition in a duct. Phil Trans R Soc Lond A 367(1888):529–544
46. Uhlmann M, Pinelli A, Kawahara G, Sekimoto A (2007) Marginally turbulent flow in a square duct. J Fluid Mech 588:153–162
47. Komminaho J, Lundbladh A, Johansson AV (1996) Very large structures in plane turbulent Couette flow. J Fluid Mech 320:259–285
48. Pirozzoli S, Bernardini M, Orlandi P (2014) Turbulence statistics in Couette flow at high Reynolds number. J Fluid Mech 758:327–343
49. Davey A (1973) On the stability of plane Couette flow to infinitesimal disturbances. J Fluid Mech 57(2):369–380
50. Romanov VA (1973) Stability of plane-parallel Couette flow. Functional analysis and its applications 7(2):137–146
51. Orszag SA, Kells LC (1980) Transition to turbulence in plane Poiseuille and plane Couette flow. J Fluid Mech 96(1):159–205
52. Lundbladh A, Johansson AV (1991) Direct simulation of turbulent spots in plane Couette flow. J Fluid Mech 229:499–516
53. Tillmark N, Alfredsson PH (1992) Experiments on transition in plane Couette flow. J Fluid Mech 235:89–102
54. Daviaud F, Hegseth J, Bergé P (1992) Subcritical transition to turbulence in plane Couette flow. Phys Rev Lett 69:2511–2514
55. Dauchot O, Daviaud F (1994) Finite-amplitude perturbation in plane Couette flow. EPL (Europhys Lett) 28(4):225
56. Shi L, Avila M, Hof B (2013) Scale invariance at the onset of turbulence in Couette flow. Phys Rev Lett 110(20):204502

57. Reichardt H (1959) Gesetzmabigkeiten der geradlinigen turbulenten Couette stromung. Max-Planck-Institut fur Stromungsforschung, 22
58. Robertson JM (1959) On turbulent plane-Couette flow. In: Proceedings of the 6th annual conference on fluid mechanics, University of Texas, Austin, TX, pp 169–182
59. Robertson JM, Johnson HF (1970) Turbulence structure in plane Couette flow. J Eng Mech Div 96(6):1171–1182
60. Aydin M, Leutheusser HJ (1979) Novel experimental facility for the study of plane Couette flow. Rev Sci Instrum 50(11):1362–1366
61. Aydin EM, Leutheusser HJ (1991) Plane Couette flow between smooth and rough walls. Exp Fluids 11(5):302–312
62. El Telbany MMM, Reynolds AJ (1982) The structure of turbulent plane Couette flow. J Fluids Eng 104(3):367–372
63. Lund KO, Bush WB (1980) Asymptotic analysis of plane turbulent Couette-Poiseuille flows. J Fluid Mech 96(1):81–104
64. Busse FH (1970) Bounds for turbulent shear flow. J Fluid Mech 41(1):219–240
65. Lee MJ, Kim J (1991) The structure of turbulence in a simulated plane Couette flow. In: Proceedings of 8th symposium turbulent shear flows, vol 5
66. Kitoh O, Nakabyashi K, Nishimura F (2005) Experimental study on mean velocity and turbulence characteristics of plane Couette flow: low-Reynolds-number effects and large longitudinal vortical structure. J Fluid Mech 539:199–227
67. Kitoh O, Umeki M (2008) Experimental study on large-scale streak structure in the core region of turbulent plane Couette flow. Phys Fluids 20(2):025107
68. Tsukahara T, Kawamura H, Shingai K (2006) Dns of turbulent Couette flow with emphasis on the large-scale structure in the core region. J Turbul 7:N19
69. Avsarkisov V, Hoyas S, Oberlack M, García-Galache JP (2014) Turbulent plane Couette flow at moderately high Reynolds number. J Fluid Mech 751:R1
70. Papavassiliou DV, Hanratty TJ (1997) Interpretation of large-scale structures observed in a turbulent plane Couette flow. Int J Heat Fluid Flow 18(1):55–69
71. Hultmark M, Vallikivi M, Bailey SCC, Smits AJ (2012) Turbulent pipe flow at extreme Reynolds numbers. Phys Rev Lett 108(9):094501
72. Hsu HW, Hsu JB, Lo W, Lin CA (2012) Large eddy simulations of turbulent Couette-Poiseuille and Couette flows inside a square duct. J Fluid Mech 702:89–101
73. Lo W, Lin CA (2006) Mean and turbulence structures of Couette-Poiseuille flows at different mean shear rates in a square duct. Phys Fluids 18(6):068103
74. Toms BA (1948) Some observations on the flow of linear polymer solutions through straight tubes at large Reynolds numbers. In: Proceedings of the 1st international congress on rheology, vol 2, pp 135–141
75. Lumley JL (1969) Drag reduction by additives. Annu Rev Fluid Mech 1(1):367–384
76. Virk PS (1975) Drag reduction fundamentals. AIChE J 21(4):625–656
77. Den Toonder JMJ, Draad AA, Kuiken GDC, Nieuwstadt FTM (1995) Degradation effects of dilute polymer solutions on turbulent drag reduction in pipe flows. Appl Sci Res 55(1):63–82
78. White CM, Mungal MG (2008) Mechanics and prediction of turbulent drag reduction with polymer additives. Annu Rev Fluid Mech 40:235–256
79. Procaccia I, L'vov VS, Benzi R (2008) Colloquium: theory of drag reduction by polymers in wall-bounded turbulence. Rev Mod Phys 80(1):225
80. Leal LG (1990) Dynamics of dilute polymer solutions. In: Structure of turbulence and drag reduction. Springer, pp 155–185
81. Graessley WW (1980) Polymer chain dimensions and the dependence of viscoelastic properties on concentration, molecular weight and solvent power. Polymer 21(3):258–262
82. Barnes HA, Hutton JF, Walters K (1989) An introduction to rheology. Elsevier, Oxford
83. Erhard P, Etling D, Muller U, Riedel U, Sreenivasan KR, Warnatz J (2010) Prandtl-essentials of fluid mechanics, vol 158. Springer Science & Business Media, London
84. Carreau PJ (1972) Rheological equations from molecular network theories. Trans Soc Rheol 16(1):99–127

85. Yasuda KY, Armstrong RC, Cohen RE (1981) Shear flow properties of concentrated solutions of linear and star branched polystyrenes. Rheol Acta 20(2):163–178
86. Metzner AB, Metzner AP (1970) Stress levels in rapid extensional flows of polymeric fluids. Rheol Acta 9(2):174–181
87. Poole RJ (2012) The deborah and weissenberg numbers. Br Soc Rheol Rheol Bull 53:32–39
88. Halliwell NA, Lewkowicz AK (1975) Investigation into the anomalous behavior of pitot tubes in dilute polymer solutions. Phys Fluids 18(12):1617–1625
89. Graham MD (2014) Drag reduction and the dynamics of turbulence in simple and complex fluids. Phys Fluids 26(10):101301
90. Warholic MD, Massah H, Hanratty TJ (1999) Influence of drag-reducing polymers on turbulence: effects of Reynolds number, concentration and mixing. Exp Fluids 27(5):461–472
91. Ptasinski PK, Nieuwstadt FTM, Van Den Brule BHAA, Hulsen MA (2001) Experiments in turbulent pipe flow with polymer additives at maximum drag reduction. Flow Turbul Combust 66(2):159–182
92. Tiederman WG (1990) The effect of dilute polymer solutions on viscous drag and turbulence structure. In: Structure of turbulence and drag reduction. Springer, pp 187–200
93. White C, Dubief Y, Klewicki J (2012) Re-examining the logarithmic dependence of the mean velocity distribution in polymer drag reduced wall-bounded flow. Phys Fluids 24(2):021701
94. Sureshkumar R, Beris AN, Handler RA (1997) Direct numerical simulation of the turbulent channel flow of a polymer solution. Phys Fluids 9(3):743–755
95. Housiadas KD, Beris AN (2003) Polymer-induced drag reduction: effects of the variations in elasticity and inertia in turbulent viscoelastic channel flow. Phys Fluids 15(8):2369–2384
96. Xi L, Graham MD (2010) Turbulent drag reduction and multistage transitions in viscoelastic minimal flow units. J Fluid Mech 647:421–452
97. Purnode B, Crochet M (1996) Flows of polymer solutions through contractions. Part 1: flows of polyacrylamide solutions through planar contractions. J Non-Newtonian Fluid Mech 65(2–3):269–289
98. Stone PA, Graham MD (2003) Polymer dynamics in a model of the turbulent buffer layer. Phys Fluids 15(5):1247–1256
99. Dubief Y, White CM, Terrapon VE, Shaqfeh ESG, Moin P, Lele SK (2004) On the coherent drag-reducing and turbulence-enhancing behaviour of polymers in wall flows. J Fluid Mech 514:271–280
100. Dubief Y, Terrapon VE, White CM, Shaqfeh ESG, Moin P, Lele SK (2005) New answers on the interaction between polymers and vortices in turbulent flows. Flow Turbul Combust 74(4):311–329
101. Shahmardi A, Zade S, Ardekani MN, Poole RJ, Lundell F, Rosti ME, Brandt L (2019) Turbulent duct flow with polymers. J Fluid Mech 859:1057–1083
102. Ptasinski PK, Boersma BJ, Nieuwstadt FTM, Hulsen MA, Van den Brule BHAA, Hunt JCR (2003) Turbulent channel flow near maximum drag reduction: simulations, experiments and mechanisms. J Fluid Mech 490:251–291
103. De Angelis E, Casciola CM, Benzi R, Piva R (2005) Homogeneous isotropic turbulence in dilute polymers. J Fluid Mech 531:1–10
104. Berman NS (1978) Drag reduction by polymers. Annu Rev Fluid Mech 10(1):47–64
105. Lumley JL (1973) Drag reduction in turbulent flow by polymer additives. J Polym Sci Macromol Rev 7:263–290
106. Tabor M, De Gennes PG (1986) A cascade theory of drag reduction. EPL (Europhy Lett) 2(7):519
107. Ryskin G (1987) Turbulent drag reduction by polymers: a quantitative theory. Phys Rev Lett 59(18):2059
108. L'vov VS, Pomyalov A, Procaccia I, Tiberkevich V (2004) Drag reduction by polymers in wall bounded turbulence. Phys Rev Lett 92(24):244503
109. Sreenivasan KR, White CM (2000) The onset of drag reduction by dilute polymer additives, and the maximum drag reduction asymptote. J Fluid Mech 409:149–164

110. Bazilevsky A, Entov V, Rozhkov A (1990) Liquid filament microrheometer and some of its applications. In: Third European rheology conference and golden jubilee meeting of the British society of rheology. Springer, pp 41–43
111. Rodd L, Scott TP, Cooper-White JJ, McKinley GH (2005) Capillary break-up rheometry of low-viscosity elastic fluids. Appl Rheol 15:12–27
112. Oliver D, Bakhtiyarov S (1983) Drag reduction in exceptionally dilute polymer solutions. J Non-Newtonian Fluid Mech 12(1):113–118
113. Metzner AB, Park MG (1964) Turbulent flow characteristics of viscoelastic fluids. J Fluid Mech 20(02):291–303
114. Darby R, Chang HFD (1984) Generalized correlation for friction loss in drag reducing polymer solutions. AIChE J 30(2):274–280
115. James DF, Yogachandran N (2006) Filament-breaking length—a measure of elasticity in extension. Rheol Acta 46(2):161–170
116. Lindner A, Vermant J, Bonn D (2003) How to obtain the elongational viscosity of dilute polymer solutions? Phys A 319:125–133
117. Zell A, Gier S, Rafai S, Wagner C (2010) Is there a relation between the relaxation time measured in CaBER experiments and the first normal stress coefficient? J Non-Newtonian Fluid Mech 165(19):1265–1274
118. Escudier MP, Presti F, Smith S (1999) Drag reduction in the turbulent pipe flow of polymers. J Non-Newtonian Fluid Mech 81(3):197–213
119. Dontula P, Pasquali M, Scriven LE, Macosko CCW (1997) Can extensional viscosity be measured with opposed-nozzle devices? Rheol Acta 36(4):429–448

Chapter 3
Experimental Methods

3.1 Experimental Arrangements and Instrumentation

In this study, experiments were conducted in a cylindrical pipe, a rectangular channel and a square duct—each a commonly encountered geometry in the study of
wall-bounded turbulence. The experimental rigs used (Figs. 3.1 and 3.2) have similar arrangements to those employed in previous research at the Fluids Engineering
laboratory at the University of Liverpool [1–3].

For the pipe flow studies, a 23 m long pipe consisting of a series of borosilicate
glass tubes with internal diameter ($2R$) of 100 mm was employed. A Dantec Dynamics stereoscopic particle image velocimetry (SPIV) system was used to obtain the
three-component velocity field over the entire cross section of the pipe at a distance
of $440R$ from the inlet. The rectangular channel consists of five 1.2 m long stainless steel sections, each of cross-sectional dimensions 25 mm $\times$ 298 mm ($2h \times W$),
followed by an 0.25 m long module fabricated with borosilicate glass side walls, to
allow for LDV measurements, and another stainless steel section of length 1.2 m,
bringing the total length to 7.45 m.

The square duct has a working-section consisting of nine stainless steel modules,
each having a length of 1.2 m and cross-sectional dimensions 80 mm $\times$ 80 mm
($2h \times 2h$). A transparent section, 150 mm in length constructed from perspex is
introduced between the eighth and ninth module to provide optical access for LDV
measurements at a distance of about $240h$ from the inlet. Since LDV is a single-point
measurement technique, the memory requirement for the data collected is relatively
small, hence it is suitable for the long-time flow statistics obtained in Chap. 5. On the
other hand, the number of frames that can be acquired in a single SPIV measurement is
limited by the internal memory of the camera. For the Couette flow studies discussed
in Appendix A, the transparent section in the square duct rig was replaced with a
module which had a moving wall.

For all rigs, efforts were made to minimise inlet effects. A cylindrical plenum-
chamber is introduced at the inlet of the pipe to reduce the degree of swirl and ensure
uniform flow. Transition sections designed to vary in cross-section from circular to

© Springer Nature Switzerland AG 2019
B. Owolabi, *Characterisation of Turbulent Duct Flows*,
Springer Theses, https://doi.org/10.1007/978-3-030-19745-2_3

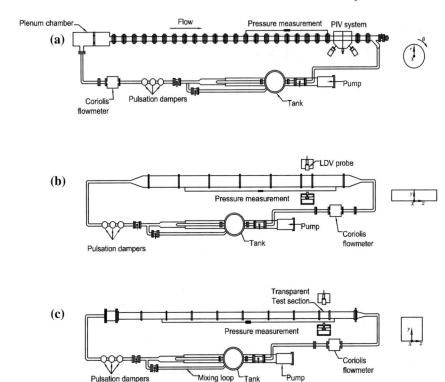

Fig. 3.1 Experimental arrangements (not to scale) of the the three parallel-shear flows: **a** pipe, **b** channel, and **c** square duct. Flow is clockwise in all three rigs. The axes systems employed are shown on the right hand side, with the streamwise/axial direction, x, pointing into the page

rectangular (or square) or vice versa are also introduced at the inlet and outlet of the rectangular and square channels to ensure smoothly varying flow. In the square duct, a honeycomb flow straightener and an additional fine mesh were also installed (these were removed whilst polymer solutions were being investigated to avoid degrading the fluids).

In all three experimental rigs, the working fluid was recirculated using progressive cavity pumps (Mono E101, with a maximum flow rate of $0.025\ \mathrm{m^3 s^{-1}}$) and pulsation dampers installed immediately after the pumps served to remove possible disturbances in the flow. Mass flow rate, density and temperature[1] were measured using Coriolis mass flow meters (Endress and Hauser Promass I, Promass F and Promass 60 in the square duct, pipe and channel rigs respectively), while pressure-drop

---

[1]A platinum resistance thermometer was used for temperature measurements in the rectangular channel.

(a)

(b)

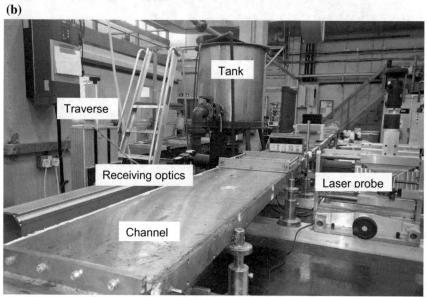

Fig. 3.2 Photographs showing the experimental setups: **a** pipe, **b** channel, **c** square duct with moving-wall test section, **d** square duct with stationary-wall test section

(c)

(d)

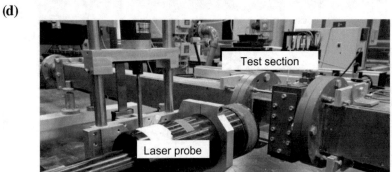

Fig. 3.2 (continued)

measurements were obtained using a Validyne differential pressure transducer (DP 15–26) over lengths of $165R$, $452h$ and $215h$ for the pipe, rectangular channel and square duct respectively. Further details on pressure drop measurements are given in Sect. 3.4.

3.2 Laser Doppler Velocimetry Measurements

Laser Doppler Velocimetry is a non-intrusive method of velocity measurement widely used in fluid dynamics research due to its high accuracy and good spatial and temporal resolutions [4]. The technique involves sending a coherent monochromatic light beam towards a target, collecting the light reflected by neutrally buoyant particles which follow the flow and obtaining the velocity of the fluid in the target area by relating it to the change in frequency of reflected light due to the Doppler effect. In this section, a brief overview of the LDV measurement principle is provided. Further details can be found in Zhang, Durst et al., Albrecht et al. [4–6] among others. In its most common implementation, two laser beams obtained by splitting a single beam are made to intersect at a target location in the flow, known as the measurement volume. This leads to the formation of an interference pattern consisting of light and dark fringes. As seeding particles in the fluid pass through the measurement volume, the laser light is scattered in all directions[2]; the scattered light (which is of varying intensity) is passed through a receiving lens into a photodetector which converts the fluctuations in light intensity into current signal fluctuations. A signal processor is then employed to determine the frequency of fluctuating current signals and hence the flow velocity.

The above explanation for the operating principle of LDV is based on the so-called fringe model. The fringe spacing (d_f) is given by:

$$d_f = \frac{\Lambda}{2sin(\varphi/2)}, \tag{3.1}$$

where φ is the angle between the two beams and Λ is the wavelength. The velocity component, V, perpendicular to the fringe pattern is given by the following expression:

$$V = \frac{d_f}{1/f_D} = f_D\left(\frac{\Lambda}{2sin(\varphi/2)}\right), \tag{3.2}$$

where the Doppler shift, f_D, is the frequency of the current signal from the photomultiplier. The parameter within brackets in Eq. 3.2 is a constant whose value is known, hence the relationship between V and f_D is linear and no calibration is required. In order to measure other velocity components, additional beam pairs are required, with all beams intersecting in a common volume.

The Doppler frequency recorded by the photomultiplier depends only on the magnitude of V, hence there is an ambiguity in the direction of the velocity measured. To correct for this, a frequency shift, f_s is usually applied to one of the laser beams,

---

[2]According to the Lorenz-Mie scattering theory, the light scattered in the forward direction is of the highest intensity [5].

using a Bragg cell. This causes the fringe pattern in the measuring volume to move at a speed, $f_s.d_f$. The frequency detected by the photomultiplier is now given by:

$$f_D = f_s + \frac{2V \sin(\varphi/2)}{\Lambda}, \tag{3.3}$$

with positive V resulting in $f_D > f_s$ and negative values yielding $f_D < f_s$. The ambiguity in direction is thus eliminated provided that $f_s > |(2V/\Lambda)sin(\varphi/2)|$. A draw back of LDV is that measurements can only be obtained for a single point at a time. The fluid and walls of the test section also have to be transparent to allow for optical access.

The LDV system employed in this study was a two-dimensional (2D) Dantec fibreflow system operated in forward-scatter mode. It consisted of a 60×10 probe, 55×12 beam expander, an Argon-Ion laser source which supplied lights of wavelength 515.5 nm (green) and 488 nm (blue) for resolving the velocity components in the streamwise and wall-normal directions (respectively x and y as defined in Fig. 3.1) and a Bragg cell which applied a frequency shift of 40 MHz. The front lens of the laser probe had a focal length of 160 mm and a beam separation distance of 51.5 mm resulting in a measuring volume of diameter $24\,\mu$m and length $150\,\mu$m in air. With this configuration, the typical data rates were around 100 Hz in non-coincidence mode and 20 Hz in coincidence mode. Signal processing was carried out using a Dantec Burst Spectrum analyser (model F50) while data acquisition and processing was done using the Dantec BSA flow software (version 2.12.00.15). Figure 3.3 illustrates the setup of the LDV system. Most of the time, particles naturally present in the fluids were employed for the measurements, but when not sufficient, Timiron seeding particles (model Supersilk MP-1005) having an average size of $5\,\mu$m were used. To correct for the effect of refraction in the z direction (see the axes systems in Fig. 3.1), the measuring volume was traversed across the transparent test section from one end to another while the rigs were filled with the working fluid, in order to obtain the linear relationship (see [5]) between the measuring volume position within the duct and probe position outside the duct. A probe movement of 0.75 mm outside the duct was found to be equivalent to a movement of 1 mm within the duct when it was filled with water or polymer solution. The procedure was repeated for the various glycerol/water and polymer solutions used. The change in refractive indices of the polymer solutions with respect to that of water was found to be negligible.

To correct for velocity bias (the phenomenon whereby high velocity particles are more frequently sampled than those of low velocities, thus bringing about a shift in the sample mean of velocities towards a higher value [5]), transit time of particles through the measurement volume was used as a weighting factor while computing the flow statistics. For the LDV measurements reported in this thesis, the difference between transit-time-weighted statistics and those obtained from an equal weighting of the data was, however, observed to be less than 1%. The expressions for the mean,

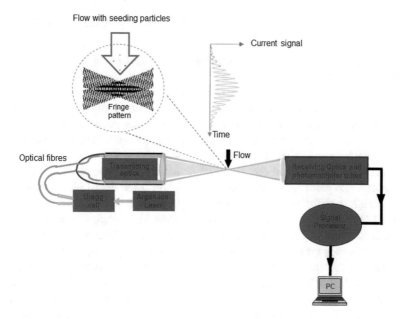

Fig. 3.3 Setup of LDV system

the root mean square velocity fluctuations, skewness and covariance are presented
in Eqs. 3.4–3.7 respectively:

$$\bar{u} = \sum_{i=1}^{N} \mathcal{F}_i u_i \tag{3.4}$$

$$u_{rms} = \sqrt{\sum_{i=1}^{N} \mathcal{F}_i (u_i - \bar{u})^2} \tag{3.5}$$

$$S = \frac{1}{u_{rms}^3} \sum_{i=1}^{N} \mathcal{F}_i (u_i - \bar{u})^3 \tag{3.6}$$

$$\overline{u'v'} = \sum_{i=1}^{N} \mathcal{F}_i (u_i - \bar{u})(v_i - \bar{v}), \tag{3.7}$$

where $\mathcal{F}_i = t_i / \sum_{i=1}^{N} t_i$, t_i is the transit time of the ith particle through the measurement volume and N is the number of samples.

3.3 Stereoscopic Particle Image Velocimetry Measurements

Particle image velocimetry is an optical measurement technique in which velocity is estimated from the displacement of tracer particles in a flow during a well defined time period [7, 8]. The region of interest is illuminated by a laser sheet and particle images are captured by a digital camera or photographic film in order to record their location. Time is marked by two short laser pulses, each typically 5–10 ns in duration [7]. Provided that the time between pulses (Δt) is small, the instantaneous velocity can be accurately determined from first order finite difference as the ratio of measured displacement and Δt. To ensure that a true estimate of the fluid velocity is obtained, the particles must be small enough to accurately follow the flow, but large enough to reflect sufficient amount of light which can be detected by a camera. Unlike LDV which is a single-point measurement technique, PIV can be used to measure the velocity field across an entire cross-section. The procedure described above allows only for the measurement of the two velocity components in the plane of the laser sheet. To obtain the 3 components of velocity, two cameras looking at the flow at different angles are required. This configuration is referred to as stereoscopic PIV and has been extensively employed in fluid dynamics research (see [9, 10]).

The SPIV system used in this study consists of a dual cavity, pulsed Nd:YAG Lee laser (LDP-100MQG) of wavelength 532 nm. The beam is transported by a light-guiding arm and optics are incorporated to produce a laser sheet of adjustable thickness, which is oriented perpendicular to the pipe axis. In this study the thickness of the sheet was set to be 1.5 mm. Two high-speed Phantom Miro 110 cameras are positioned on opposite sides of the laser light sheet such that the optical axes of the lens are at 45° angles to it. The cameras can acquire images at a rate of up to 1630 frames per second at full resolution of 1280×1280 pixels, and have a pixel pitch of $10 \mu m$ and a pixel depth of 12 bits/pixel. They are fitted with a 105 mm Sigma lens with a minimum f-number of 2.8 (more than enough to ensure that the field of view covers the entire pipe cross-section). To ensure that all particles in the measurement/object plane (the centre plane of the laser sheet) are in focus, the plane of the image sensor in each camera must be positioned to coincide with the image plane. The measurement, lens, and image planes also have to intersect at a point [7]. This requirement known as the Scheimpflug condition was satisfied by installing the cameras on Scheimpflug mounts which allowed the camera body to be tilted relative to the lens.

The dual frame mode in which images exposed by the laser pulses at time t and $t + \Delta t$ are recorded on two separate frames is employed. The image plane in each camera is divided into small regions called interrogation windows, and from standard space-time cross correlation between the two frames, the apparent[3] two-component displacement vectors are determined. The two-component vectors from both cameras are then back projected (dewarped) onto the measurement plane and interpolated on a common grid, and the resulting vectors are combined to reconstruct the three-component displacements and hence the 3D velocity field. For successful

---

[3]The displacement observed by each camera is that perpendicular to the viewing direction [9].

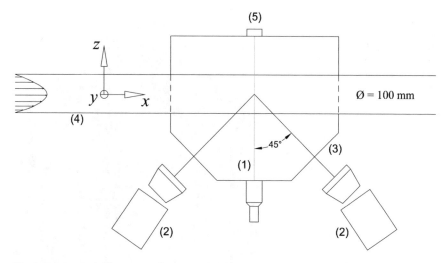

Fig. 3.4 Setup of SPIV system. (1) laser sheet; (2) high-speed camera on Scheimpflug mount; (3) water-filled prism; (4) glass pipe; (5) beam dump

dewarping and reconstruction of the vectors, accurate calibration of the SPIV system is required. This allows for a mapping function between image and object plane to be determined. In this study, a multi-level double sided calibration target with dots located on three parallel planes, 3 mm apart was employed, the middle plane coinciding with the measurement plane (set as the zero position in the streamwise direction). With this arrangement, both in-plane and out-of-plane images could be taken without having to traverse the target. For the SPIV data presented in this thesis, the calibration was carried out while the rig was filled with the working fluid. Hence errors due to changes in optical properties with the addition of polymers (in the investigation of polymer drag reduction, for example) are negligible.

A schematic of the SPIV system is shown in Fig. 3.4. The pipe measurement section is enclosed in a water-filled prism to provide undistorted optical access for the cameras. The flow was seeded with $10\,\mu$m silver-coated hollow glass spheres at a concentration of 5 g/m^3 of fluid. Mirsepassi and Rankin [11] showed that at such concentrations, polymer/particle interaction effects are not significant. The velocity data were obtained at sampling rates of up to 619 Hz and analysed using the Dynamic studio software (version 4.10) from Dantec.

3.4   Pressure Drop Measurements and Determination of Wall Shear Stress

Figure 3.5 shows the Validyne differential pressure transducer (DP 15) used in this study for mean pressure drop ($\overline{\Delta p}$) measurements. The device consists of a replace-

Fig. 3.5 Validyne differential pressure transducer (DP-15): (1) pressure port, (2) diaphragm, (3) bleed screw, (4) electronics housing, (5) electrical connector

able diaphragm clamped between two stainless steel blocks, each having an embedded inductance coil. At zero flow, the diaphragm is undeflected. However, when a pressure drop is applied through the two pressure ports machined into the steel blocks, the diaphragm bends towards the low pressure side, and a voltage signal, which varies linearly with the pressure difference, is generated. A Validyne CD223 carrier demodulator supplied with the transducer was used for signal conditioning. For measurements in the circular pipe, the voltage output from the transducer was sampled using a LabJack analogue-to-digital converter (ADC) at a rate of 100 Hz. However, in the square and rectangular channels, the ADC in-built into the burst spectrum analyser was employed, thus allowing for pressure drop readings to be synchronised with Doppler bursts from the LDV system. The Validyne pressure transducer is quoted to have an accuracy of ±0.25% full scale (±2.15 Pa). It was calibrated periodically using air, against a Baratron high-precision differential transducer manufactured by MKS Instruments, and a diaphragm with a full scale range of 0.86 kPa was used. Figure 3.6 shows a typical calibration curve.

In taking pressure drop measurements, each of the two ports on the pressure transducer was connected to a pressure tapping in an experimental rig, using transparent vinyl tubing. Special care was taken to eliminate bubbles. This was achieved by

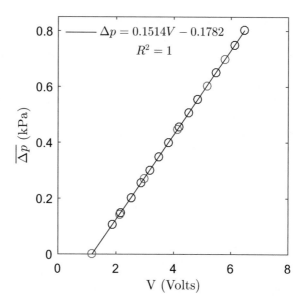

Fig. 3.6 Typical calibration curve for pressure transducer. ○, increasing pressure drop; ○, decreasing pressure drop

bleeding the transducer (through the bleed screw; see Fig. 3.5) at low flow rates before taking measurements. Similarly, the pressure drop reading at zero flow was taken before and after every experiment in order to account for any drift.

From a mean pressure drop $(\overline{\Delta p})$ measurement, the average value of the shear stress over the perimeter of a duct was determined from momentum balance as:

$$\bar{\tau}_w = \frac{\overline{\Delta p} A}{\ell q}, \tag{3.8}$$

where A is the cross-sectional area of the duct, ℓ is the length over which $\overline{\Delta p}$ occurs and q, the wetted perimeter. For a large aspect ratio channel ($W >> H$, where W is the channel width and H, the height), Eq. 3.8 simplifies to

$$\bar{\tau}_w = \frac{\overline{\Delta p} H}{2\ell}. \tag{3.9}$$

Turbulence leads to an increase in the wall-shear-stress and hence the drag, compared to that in a laminar flow. The skin friction coefficient, f given by (see e.g. [12])

$$f = \frac{\bar{\tau}_w}{\frac{1}{2}\rho U_b^2}, \tag{3.10}$$

where U_b is the bulk or average velocity along the axis of a duct, is the non-dimensional shear stress at the wall. In the case of a pipe, $\bar{\tau}_w$ is identical to the local stress value. For non-circular ducts however, this is not the case, as the wall shear stress is non-uniformly distributed over the walls; hence the local wall shear stress in the square and rectangular ducts were determined by taking LDV measurements of the velocity profile in the viscous sublayer, obtaining the gradient of the resulting linear plot and applying the expression: $\bar{\tau} = \mu \frac{d\bar{u}}{dy}$, where $\bar{u}$ is the mean streamwise velocity. Unlike in hot-wire anemometry, this technique is free from the so-called wall effect since LDV is non intrusive. The very small measuring volume minimises any potential errors due to velocity gradient broadening. A detailed discussion on wall shear stress determination using the near-wall velocity gradient method is given by Hutchins and Choi [13].

3.5   Rheological Measurements

The working fluids employed in this study include water and glycerol/water solutions (Newtonian) as well as polyacrylamide solutions of various concentrations (non-Newtonian). Polyacrylamide, a flexible polymer, is known to be a good drag-reducing agent with skin friction reduction of up to about 75% recorded in the literature [14]. Two grades of polyacrylamide solutions having different molecular weights with concentrations ranging from 150 to 350 parts per million (ppm) were studied: FloPAM AN934SH ("PAA") and Separan AP273E ("Separan") both supplied by Floreger. Intrinsic viscosity measurements $[\eta]$ (see [15]) for each grade of polymer ($[\eta]$ = 4400 ml/g for PAA and 3400 ml/g for Separan) suggest critical overlap concentrations (c^*) of 225 ppm and 300 ppm for PAA and Separan respectively. Thus all concentrations are on the order of c^* and are therefore best considered semi dilute rather than truly dilute [16]. (NB. Although there must exist some polydispersity within these polymers, it was not possible to measure their molecular weights using gel phase chromotography (GPC) due to the occurrence of viscous fingering in the GPC column).

The polymeric fluids exhibited both shear-thinning and visco-elastic behaviour. Their properties under shear and extensional deformations were measured using an Anton Paar MCR302 controlled-stress torsional rheometer and a Capillary Breakup Extensional Rheometer (CaBER) respectively. For water, the correlation of Korson et al. [17] was used to determine the shear viscosity at different temperatures:

$$\mu(T) = \mu_{20}.10^{([A(20-T)-B(T-20)^2]/[T+C])}, \tag{3.11}$$

where $\mu_{20} = 1.002 \times 10^3$ Pa.s is the viscosity at a temperature of 20 °C, A and B are constants with values of 1.1709 and 0.001827 respectively, $C = 89.93$ °C and T is the temperature in °C. Equation 3.11 is valid for temperatures ranging from 10 °C to 70 °C (experiments were conducted at temperatures which fall within this range). However, for glycerol/water solutions the shear viscosity was measured using the Anton Paar MCR302 controlled-stress torsional rheometer, also.

3.5.1 Shear Rheology Measurements

For the shear viscosity measurements, a cone and plate geometry was employed (see Fig. 3.7) due to its ability to ensure a constant shear rate across an entire fluid sample and its suitability for the low to medium shear rates encountered in this study. The fluid sample is loaded between a fixed plate and a cone which is attached to the spindle of a driving motor. The instrument applies a torque, T, thus setting up a shear flow in the fluid. The shear rate, $\dot{\gamma}$, is given by

$$\dot{\gamma} = \frac{\omega}{\theta},\tag{3.12}$$

where ω and θ represent the angular velocity and cone angles respectively (a stainless steel cone with $\theta = 2°$ was used for the measurements). The shear stress is then obtained from

$$\tau_{xy} = \frac{3T}{2\pi R^3},\tag{3.13}$$

where R is the cone radius (which in this study was 30 mm), and the apparent shear viscosity is determined from the expression, $\tau_{xy} = \mu\dot{\gamma}$. Hence the output from such measurements is a plot of shear viscosity against shear rate. During the measurements, temperature is controlled to within $\pm 0.1\,°C$ by a Peltier system incorporated into the plate. Typical shear viscosity plots are shown in Fig. 3.8. The data are for freshly prepared PAA of different concentrations and they have been fitted with the Carreau-Yasuda model ([18, 19], see Eq. 2.15). The model parameters are given in Table 3.1.

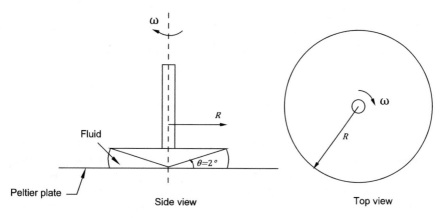

Fig. 3.7 Shear viscosity measurement using a cone and plate geometry

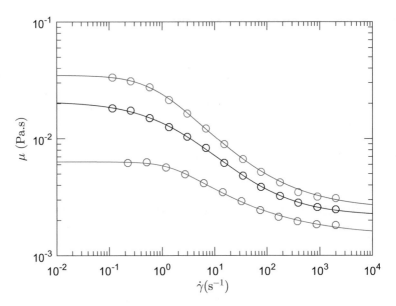

Fig. 3.8 Shear viscosity plots for different concentrations of freshly prepared PAA at a temperature of 20 °C. ○, 350 ppm; ○, 250 ppm; ○, 150 ppm. Solid lines are the corresponding Carreau-Yasuda fits

Table 3.1 Carreau-Yasuda model parameters for the different concentrations of PAA shown in Fig. 3.8

Concentration (ppm)	m	K_{cy} (s)	μ_∞ (Pa.s)	μ_0 (Pa.s)	n
150	1.654	0.675	0.00148	0.00633	0.401
250	0.666	0.603	0.00218	0.0208	0.602
350	1.17	1.419	0.00250	0.0348	0.523

3.5.2 Extensional Rheology Measurements

In this study, polymer relaxation times were measured using a CaBER. The instrument comprises two circular stainless steel platens, 4 mm in diameter (D_o), with an initial separation of ≈2 mm (see Fig. 3.9). A small sample of each solution was loaded between the platens using a syringe (without a needle to minimise the shear) to form a cylindrical sample. A rapid axial step strain was imposed (≈50 ms) until a final height (≈9 mm) was reached and an unstable filament formed. Subsequently, the sample filament breaks up under the combined action of capillary and extensional viscoelastic forces. The diameter (D_{mid}) of the filament was observed as a function of time using the laser micrometer (resolution 10 microns) incorporated in the device. Although the filament diameter data can be post-processed into an (apparent) extensional viscosity, the standard method [20] to quantify extensional

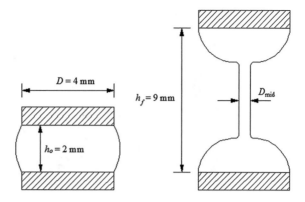

Fig. 3.9 CaBER geometry with fluid sample **a** at rest and **b** undergoing thinning

effects is via an exponential fit to the filament diameter as a function of time, t, in the elasto-capillary regime (where the decrease in diameter follows Eq. 3.14). From this fit, a characteristic relaxation time, λ_c (more correctly a characteristic time for extensional stress growth) can be obtained.

$$\frac{D_{mid}(t)}{D_o} = \left(\frac{GD_o}{4\sigma}\right)^{1/3} e^{-t/3\lambda_c} \tag{3.14}$$

Figure 3.10 shows typical plots of filament diameter against time obtained from CaBER measurement. In Eq. 3.14, G and σ represent the elastic modulus and surface tension of the filament respectively. Their values need not be known, as λ_c can be determined from the slope of the line fitted to the data when plotted in semi-log form.

3.6 Measurement Uncertainties

In this section, a discussion on the sources of error and the quantification of uncertainties associated with the various measurements recorded in this thesis is provided. For velocity data obtained using LDV, signal broadening due to instrument noise, existence of a velocity gradient in the measurement volume and finite transit times of seeding particles can significantly affect measurement accuracy [21]. As earlier explained, velocity bias also plays a role in the overall accuracy of the data. Furthermore, there is an uncertainty in the determination of the beam angle, φ in Eq. 3.2 (estimated to be about 1.5% in this thesis). Considering all these sources of error, it is estimated that there is an uncertainty of about 3% in the mean velocity data and 8% in the turbulence intensity data.

A detailed discussion on the sources of error in SPIV measurements can be found in Van Doorne and Westerweel [9]. These include correlation noise, peak-locking, and misregistration due to erroneous mapping between image and measurement plane, among others. The accuracy of the SPIV measurements presented in this

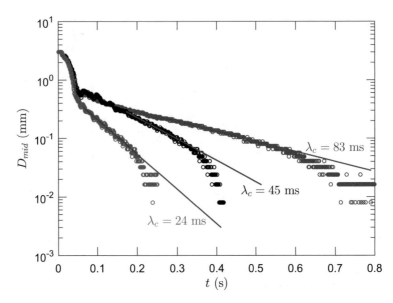

Fig. 3.10 Diameter versus time data from CaBER for various concentrations of freshly prepared PAA: ○ , 350 ppm; ○ , 250 ppm; ○ , 150 ppm. The solid lines are linear fits from whose slopes, the relaxation times are determined

thesis was assessed by comparing the Newtonian velocity profile with the DNS data of El Khoury et al. [22] at $Re_\tau = 1000$. The measurements are within 2% of the DNS data.

As earlier pointed out, the pressure transducer is quoted to have an accuracy of ± 0.125 Pa. Since the pressure drop is proportional to the length, l, over which the measurement is taken, the % error can be reduced by taking the measurements over long l. In the cylindrical pipe, rectangular channel and square duct, l was taken to be $165R$, $452h$ and $215h$ respectively.

In estimating the relaxation times of polymer solutions, the CaBER measurements were repeated at least five times for every liquid sample and the average values from the measurements were used. For freshly prepared polymer solution at the concentrations considered, the uncertainty in the data was found to be less than 10%. However for degraded polymer solutions at relaxation times less than 5 ms, up to 50% uncertainty was observed. The shear viscosity measurements obtained from the torsional rheometer have an accuracy of ± 2%.

Error bars have been included in various plots in this thesis to clearly show the uncertainties in the data presented.

References

1. Dennis DJC, Sogaro FM (2014) Distinct organizational states of fully developed turbulent pipe flow. Phys Rev Lett 113(23):234501
2. Whalley R, Park JS, Kushwaha A, Dennis DJC, Graham MD, Poole RJ (2017) Low-drag events in transitional wall-bounded turbulence. Phys Rev Fluids
3. Escudier M, Smith S (2001) Fully developed turbulent flow of non-Newtonian liquids through a square duct. Proc R Soc A 457(2008):911–936
4. Durst F, Melling A, Whitelaw JH (1981) Principles and practice of laser Doppler anemometry. Academic Press, London
5. Zhang Z (2010) LDA application methods: laser Doppler anemometry for fluid dynamics. Springer Science & Business Media, New York
6. Albrecht HE, Damaschke N, Borys M, Tropea C (2013) Laser Doppler and phase Doppler measurement techniques. Springer Science & Business Media, New York
7. Adrian RJ, Westerweel J (2011) Particle image velocimetry. Cambridge University Press, New York
8. Raffel M, Willert CE, Kompenhans J (2007) Particle image velocimetry: a practical guide. Springer Science & Business Media, New York
9. Van Doorne CWH, Westerweel J (2007) Measurement of laminar, transitional and turbulent pipe flow using stereoscopic-piv. Exp Fluids 42(2):259–279
10. Prasad AK (2000) Stereoscopic particle image velocimetry. Exp Fluids 29(2):103–116
11. Mirsepassi A, Rankin DD (2014) Particle image velocimetry in viscoelastic fluids and particle interaction effects. Exp Fluids 55(1):1641
12. White FM (2006) Viscous fluid flow, 3rd edn. McGraw-Hill, New York
13. Hutchins N, Choi K-S (2002) Accurate measurements of local skin friction coefficient using hot-wire anemometry. Prog Aerosp Sci 38(4):421–446
14. Escudier MP, Nickson AK, Poole RJ (2009) Turbulent flow of viscoelastic shear-thinning liquids through a rectangular duct: Quantification of turbulence anisotropy. J Non-Newtonian Fluid Mech 160(1):2–10
15. Poole RJ (2016) Elastic instabilities in parallel shear flows of a viscoelastic shear-thinning liquid. Phys Rev Fluids 1(4):041301
16. Clasen C, Plog J, Kulicke W-M, Owens M, Macosko C, Scriven L, Verani M, McKinley GH (2006) How dilute are dilute solutions in extensional flows? J Rheol 50(6):849–881
17. Korson L, Drost-Hansen W, Millero FJ (1969) Viscosity of water at various temperatures. J Phys Chem 73(1):34–39
18. Carreau PJ (1972) Rheological equations from molecular network theories. Trans Soc Rheol 16(1):99–127
19. Yasuda KY, Armstrong RC, Cohen RE (1981) Shear flow properties of concentrated solutions of linear and star branched polystyrenes. Rheol Acta 20(2):163–178
20. Rodd L, Scott TP, Cooper-White JJ, McKinley GH (2005) Capillary break-up rheometry of low-viscosity elastic fluids. Appl Rheol 15:12–27
21. Poole RJ (2002) Turbulent flow of Newtonian and non-Newtonian liquids through sudden expansions. PhD thesis, University of Liverpool, United Kingdom
22. El Khoury GK, Schlatter P, Noorani A, Fischer PF, Brethouwer G, Johansson AV (2013) Direct numerical simulation of turbulent pipe flow at moderately high Reynolds numbers. Flow Turbul Combust 91(3):475–495

Chapter 4
Numerical Methods

4.1 Introduction

In this chapter, details of the techniques employed in the direct numerical simulation of turbulent duct flows are presented. The finite-volume method, in which the governing flow equations are integrated over small control volumes in the problem domain, was adopted. This technique guarantees both global and local conservation of momentum and mass and is widely used in computational fluid dynamics research (see e.g. [1–3]). The turbulent simulations are those of purely wall-driven Newtonian flows in a square duct. In Sect. 4.2, the grid generation method is discussed while the procedure for discretising the governing transport equations in space and time is given in Sect. 4.3 and Sect. 4.4 respectively. The chapter is concluded by a discussion of the boundary conditions employed and a brief description of the DNS code used. Further details on this code and the numerical techniques described in this chapter can be found in Hsu [4]. In this thesis, laminar flow simulations were also carried out; however, the well-known commercial computational fluid dynamics (CFD) software, ANSYS Fluent (also a finite volume code) was used for these. As this approach is rather standard, further brief details on the problem setup in Fluent are simply provided in Chaps. 5 and 6, where the results of those computations are discussed.

4.2 Mesh Generation

Grid generation is a very important stage in any numerical simulation of fluid flows, as it is over the mesh created that the discretised governing equations will be solved. Hence to ensure that a correct solution is obtained, the grid must be generated in such a way that the flow physics is accurately captured. In turbulent wall-bounded flows, large gradients of the flow properties are observed close to a boundary. To capture the fine turbulence structure in the near-wall region, non-uniform grids have been

© Springer Nature Switzerland AG 2019
B. Owolabi, *Characterisation of Turbulent Duct Flows*,
Springer Theses, https://doi.org/10.1007/978-3-030-19745-2_4

employed in the cross-sectional plane of the square duct, the spacing increasing with distance from the wall. In the turbulent flow simulations of Chap. 6 the spacing of the grid closest to the wall ranges from 0.04 to 0.13 wall units (see Table 6.1 for details). Since fully developed flows are being considered, there is a zero streamwise/axial gradient of the mean flow properties, hence the mesh is uniformly spaced in the streamwise direction. An additional requirement in turbulent flow computations is that the grid spacing be of the order of the local Kolmogorov length scales, and this requirement was ensured in all of the simulations (see Chap. 6 for details). The stretching in the cross-sectional plane is governed by hyperbolic tangent functions given by

$$y = \xi \left[1 - \frac{\tanh \gamma(\xi - \zeta)}{\tanh \gamma \xi} \right] \qquad (4.1)$$

$$z = \xi \left[1 - \frac{\tanh \gamma(\xi - \chi)}{\tanh \gamma \xi} \right], \qquad (4.2)$$

where γ determines the degree of compression near the boundary (values used in this study range between 4 and 4.5), ξ controls the symmetry of the resulting grid (it was set to half of the duct's diameter to ensure symmetry about the wall bisectors), ζ and χ are the coordinates of a uniformly spaced mesh with one to one correspondence to y and z respectively (the grid points in the wall-normal directions). Figure 4.1 shows a typical non-uniform mesh with 96×96 cells in the $y - z$ plane.

In this study, a staggered grid arrangement is employed. Values of pressure are stored at the centroid of the control volumes, while velocity components are stored at the faces. This is equivalent to having separate control volumes for each velocity

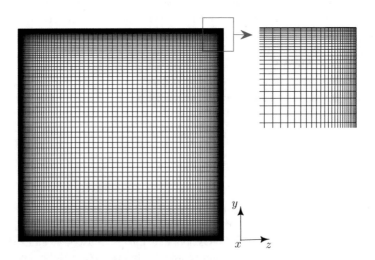

Fig. 4.1 A typical clustered mesh. y and z are the wall normal directions, while x is the streamwise direction (into the page)

component, shifted relative to those of pressure. Further details are given in Sect. 4.3. The use of a staggered mesh ensures a strong coupling between pressure and velocity and eliminates checker-board errors [1]. However, it is more challenging to keep track of the flow variables, since they are stored in different locations.

4.3 Spatial Discretisation

The governing partial differential equations being considered are those expressing the conservation of mass and momentum, which in Einstein's notation, are:

$$\frac{\partial u_i}{\partial x_i} = 0, \tag{4.3}$$

$$\frac{\partial \rho u_i}{\partial t} = -\frac{\partial \rho u_j u_i}{\partial x_j} + \frac{\partial}{\partial x_j}\mu\left(\frac{\partial u_i}{\partial x_j}\right) - \frac{\partial p}{\partial x_i}. \tag{4.4}$$

The streamwise direction is x (corresponding to $i = 1$) while y and z (corresponding to $i = 2$ and 3) are the transverse and spanwise directions respectively, with u, v and w being the corresponding velocity components (i.e u_1, u_2 and u_3 respectively). p and t represent pressure and time respectively. In the DNS, the effect of Coriolis and other body forces were not considered, hence the Coriolis term is not included in Eq. 4.4.

First, the treatment of the momentum equation is presented. Equation 4.5 is the conservation form of Eq. 4.4, written for the ith velocity component. This is integrated over the control volume, V, for that velocity component (see Eq. 4.6, where $dV = dxdydz$).

$$\frac{\partial \rho u_i}{\partial t} = -\left[\frac{\partial}{\partial x}\left(\rho u u_i\right) + \frac{\partial}{\partial y}\left(\rho v u_i\right) + \frac{\partial}{\partial z}\left(\rho w u_i\right)\right] + \left[\frac{\partial}{\partial x}\left(\mu\frac{\partial u_i}{\partial x}\right) + \frac{\partial}{\partial y}\left(\mu\frac{\partial u_i}{\partial y}\right) + \frac{\partial}{\partial z}\left(\mu\frac{\partial u_i}{\partial z}\right)\right] - \frac{\partial p}{\partial x_i} \tag{4.5}$$

$$\iiint_V \frac{\partial \rho u_i}{\partial t}dV = -\iiint_V\left[\frac{\partial}{\partial x}\left(\rho u u_i\right) + \frac{\partial}{\partial y}\left(\rho v u_i\right) + \frac{\partial}{\partial z}\left(\rho w u_i\right)\right]dV + \iiint_V\left[\frac{\partial}{\partial x}\left(\mu\frac{\partial u_i}{\partial x}\right) + \frac{\partial}{\partial y}\left(\mu\frac{\partial u_i}{\partial y}\right) + \frac{\partial}{\partial z}\left(\mu\frac{\partial u_i}{\partial z}\right)\right]dV - \iiint_V \frac{\partial p}{\partial x_i}dV \tag{4.6}$$

Gauss's divergence theorem is applied to convert the volume integrals for the convection and diffusion terms to surface integrals, while the pressure gradient and the term involving a time derivative are converted to discrete forms by applying the mean value theorem. The resulting discretised equation is given by

$$\frac{\partial \rho u_i}{\partial t} \Delta x \Delta y \Delta z = -\left[\left(\rho u u_i|_e - \rho u u_i|_w\right)\Delta y \Delta z + \left(\rho v u_i|_n - \rho v u_i|_s\right)\Delta x \Delta z + \right.$$
$$\left.\left(\rho w u_i|_t - \rho w u_i|_b\right)\Delta x \Delta y\right] + \left[\left(\mu\frac{\delta u_i}{\delta x}\right)_e - \left(\mu\frac{\delta u_i}{\delta x}\right)_w\right]\Delta y \Delta z +$$
$$\left[\left(\mu\frac{\delta u_i}{\delta y}\right)_n - \left(\mu\frac{\delta u_i}{\delta y}\right)_s\right]\Delta x \Delta z + \left[\left(\mu\frac{\delta u_i}{\delta z}\right)_t - \left(\mu\frac{\delta u_i}{\delta z}\right)_b\right]\Delta x \Delta y - \frac{\delta p}{\delta x_i}\Delta x \Delta y \Delta z.$$

$$(4.7)$$

$\delta/\delta x_i$ represents a spatial derivative discretised using the second-order central difference scheme, for example,

$$\left(\frac{\delta u}{\delta x}\right)_e = \frac{u_E - u_P}{\Delta x}, \tag{4.8}$$

and e, w, n, s, t and b represent the east, west, north, south, top and bottom faces of the control volume. P is the cell centroid and E, the centroid of the neighbouring control volume (see Fig. 4.2). For simplicity the discretised expressions for a uniform grid are presented. As previously stated, a mesh which is non-uniform in the $y - z$ plane is employed for the simulations. Hence to maintain second order accuracy, the solution domain is mapped onto a uniform grid ($\zeta - \chi$) on which the derivatives of a flow variable, ϕ are computed, before transforming back to a non uniform grid using Eqs. 4.9 and 4.10.

$$\frac{\partial \phi}{\partial z} = \frac{\partial \phi}{\partial \chi} \cdot \frac{\partial \chi}{\partial z} \tag{4.9}$$

$$\frac{\partial \phi}{\partial y} = \frac{\partial \phi}{\partial \zeta} \cdot \frac{\partial \zeta}{\partial y} \tag{4.10}$$

The values of $\partial \chi/\partial z$ and $\partial \zeta/\partial y$ are known exactly from Eqs. 4.1 and 4.2.

Because a staggered grid is used, the value of p is known at the faces of the u, v and w control volumes (see Fig. 4.2). Values of other variables, ϕ, at the faces are obtained by linear interpolation between the two nearest nodes. For example,

$$\phi_e = \phi_E \Gamma_e + \phi_P(1 - \Gamma_e), \tag{4.11}$$

where $\Gamma_e = (x_e - x_P)/(x_E - x_P)$. This reduces to Eq. 4.12 for the case of a uniform grid:

$$\phi_e = \frac{\phi_E + \phi_P}{2}. \tag{4.12}$$

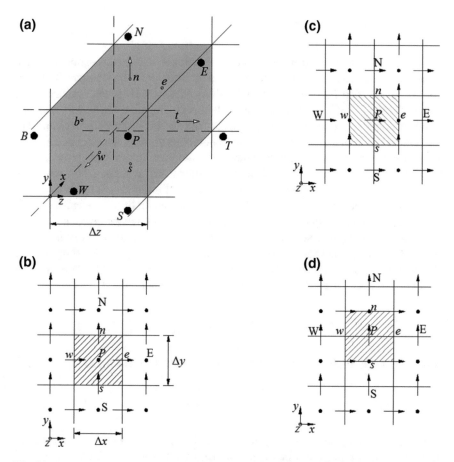

Fig. 4.2 Staggered grid arrangement. **a** is a typical control volume for p in 3D. The shaded cells in **b**, **c** and **d** are control volumes for p, u and v in the $x - y$ plane. P represents the centroid of a control volume, and E, W, N, S, T, B, those of the neighbouring control volumes, while e, w, n, s, t, b represent the faces. Velocity components are represented by arrows, while pressure is represented by filled circles

Equation 4.7 can be written in compact form as:

$$\frac{\partial \rho u_i}{\partial t} = -C_i + D_i - \frac{\delta p}{\delta x_i}, \tag{4.13}$$

where C_i, D_i and $\delta p / \delta x_i$ represent the discretised convection, diffusion and pressure gradient terms.

4.4 Temporal Discretisation

Equation 4.7 is discretised in time using a fractional step method similar to that employed by Kim and Moin [5]. Convection and diffusion in the wall-parallel direction are advanced in time using the Adams-Bashforth scheme (a second-order explicit scheme; see [1]), while diffussion in the wall normal directions are treated using the Crank-Nicolson method (an implicit scheme with second-order accuracy; see [1]). As pointed out by Kim and Moin [5], the implicit treatment of the viscous diffusion terms ensures numerical stability over the entire computational domain, provided that the CFL condition is satisfied. This however results in an increase in the complexity of the computations due to the need to invert a matrix. Further details on the temporal discretisation procedure employed in the DNS code used for the computations in this thesis can be found in Hsu et al. [6]. A brief summary is however given here.

In implementing the fractional step method, the velocity and pressure fields at the current time step, n, are assumed to be known. Values at the $(n + 1)$th time step are obtained through the following four steps. In the first step, Eq. 4.14 is solved on all grid cells, to obtain the velocity field, u_i^*, at an intermediate time $*$.

$$\rho \frac{u_i^* - u_i^n}{\Delta t} = -\left(\frac{3}{2} C_i^n - \frac{1}{2} C_i^{n-1} \right) + \left(\frac{3}{2} D_{i,p}^n - \frac{1}{2} D_{i,p}^{n-1} \right) + \frac{1}{2} \left(D_{i,n}^n + D_{i,n}^* \right) - \frac{\delta p^n}{\delta x_i} \tag{4.14}$$

Next, u_i^{**} at a second intermediate time $**$ is obtained from Eq. 4.15.

$$\rho \frac{u_i^{**} - u_i^*}{\Delta t} = \frac{\delta p^n}{\delta x_i}. \tag{4.15}$$

In the third step, the Poisson equation (4.16) is solved to obtain the pressure field at the $(n + 1)$th time step. Finally, the velocity field at the $(n + 1)$th time step is evaluated from Eq. 4.17.

$$\frac{\delta^2 p^{n+1}}{\delta x^2} = \frac{\rho}{\Delta t} \frac{\delta u_i^{**}}{\delta x_i}. \tag{4.16}$$

$$\rho \frac{u_i^{n+1} - u_i^{**}}{\Delta t} = -\frac{\delta p^{n+1}}{\delta x_i} \tag{4.17}$$

By taking the divergence of (4.17), it can be shown that the pressure field obtained in the step 3 satisfies the continuity equation. Since the flows considered are incompressible, momentum transport depends only on the pressure gradient and the actual values of pressure are irrelevant.

4.4.1 Poisson Equation for Pressure

In this section, the method for discretising and solving the pressure Poisson equation is explained. Equation 4.16 is integrated over the control volume for pressure to obtain the following:

$$\iiint_V \frac{\partial^2 p}{\partial x_i^2} \, d\Psi = \iiint_V \frac{\rho}{\Delta t} \frac{\partial u_i^{**}}{\partial x_i} \, d\Psi \tag{4.18}$$

$$\iiint_V \frac{\partial^2 p}{\partial x^2} + \frac{\partial^2 p}{\partial y^2} + \frac{\partial^2 p}{\partial z^2} \, d\Psi = \iiint_V \frac{\rho}{\Delta t} \left(\frac{\partial u^{**}}{\partial x} \frac{\partial v^{**}}{\partial y} \frac{\partial w^{**}}{\partial z} \right) d\Psi \tag{4.19}$$

Application of Gauss's divergence theorem results in

$$\left[\left(\frac{\delta p}{\delta x} \right)_e - \left(\frac{\delta p}{\delta x} \right)_w \right] \Delta y \Delta z + \left[\left(\frac{\delta p}{\delta y} \right)_n - \left(\frac{\delta p}{\delta y} \right)_s \right] \Delta x \Delta z +$$

$$\left[\left(\frac{\delta p}{\delta z} \right)_t - \left(\frac{\delta p}{\delta z} \right)_b \right] \Delta x \Delta y = \frac{\rho}{\Delta t} \left[(u_e^{**} - u_w^{**}) \Delta y \Delta z + (v_n^{**} - v_s^{**}) \Delta x \Delta z + \right.$$

$$\left. (w_t^{**} - w_b^{**}) \Delta x \Delta y \right]. \tag{4.20}$$

As explained in Sect. 4.3, $\delta/\delta x_i$ represents a spatial derivative discretised using the central difference scheme, for example,

$$\left(\frac{\delta p}{\delta x} \right)_e = \frac{p_E - p_P}{\Delta x}, \tag{4.21}$$

where p_E, p_W, etc. are, respectively, the values of pressure at the east and west nodes etc. (see Fig. 4.2). Substituting the expressions for $\delta p/\delta x_i$ into Eq. 4.20, the resulting equation can be written in the following form:

$$A_P p_P = A_E p_E + A_W p_W + A_N p_N + A_S p_S + A_T p_T + A_B p_B + b, \tag{4.22}$$

where A_P, A_W, etc. are constants and

$$b = -\frac{\rho}{\Delta t} \left[(u_e^{**} - u_w^{**}) \Delta y \Delta z + (v_n^{**} - v_s^{**}) \Delta x \Delta z + (w_t^{**} - w_b^{**}) \Delta x \Delta y \right]. \tag{4.23}$$

When Eq. 4.22 is written for all the pressure control volumes, the system of Eq. 4.24 (written in matrix notation) is obtained:

$$\mathbf{Ap} = \mathbf{b} \tag{4.24}$$

An FFT-based Poisson solver was employed in this study. The three dimensional Poisson equation (4.22) was reduced, using fast Fourier transform, to uncoupled two-dimensional algebraic equations which were solved by LU decomposition. Further details on the solver can be found in the thesis of Hsu [4]. The solution of the Poisson equation is very time consuming, taking up to 60% of total simulation time [4].

4.5 Boundary Conditions

The choice of boundary conditions is a very important consideration in the numerical simulation of any partial differential equation. The incompressible Navier-Stokes equations show elliptic and parabolic behaviour in space and time, respectively, hence boundary conditions have to be specified over the entire computational domain and initial conditions also have to be defined. For the momentum equations, the no slip condition (a Dirichlet boundary condition) was imposed at the walls. This implies that the fluid velocity at a wall is equal to the wall velocity. In the streamwise direction, a periodic boundary condition was employed. In solving the Poisson equation for pressure, a Neumann boundary condition given by Eq. 4.25 was prescribed at the walls:

$$\frac{\partial p}{\partial n} = 0, \tag{4.25}$$

where n is the normal direction to a wall. Equation 4.25 arises from the impermeability condition.

4.6 DNS Code

The DNS programme was developed by Hsu [4], but some modifications were introduced in this study (see Chap. 6 for further details). It is a parallel code written in Fortran 77 using the Message Passing Interface (MPI) library. The code was validated by simulating the two-dimensional laminar decaying vortex problem (Taylor-Green flow), for which an analytical solution exists (see [4, 6, 7]).

The code was also benchmarked against the DNS results of Gavrilakis [8] for purely pressure-driven flow in a square duct, with excellent agreement between the two and it has been used previously in the simulation of Couette-Poiseuille flows in a square duct (see [9]).

References

1. Ferziger JH, Peric M (2012) Computational methods for fluid dynamics, 3rd edn. Springer Science & Business Media, London
2. John D, Anderson JR (1995) Computational fluid dynamics: the basics with applications
3. Patankar S (1980) Numerical heat transfer and fluid flow. Hemisphere Publishing Corporation, Washington DC
4. Hsu HW (2012) Investigation of turbulent Couette-Poiseuille and Couette flows inside a square duct. PhD thesis, National Tsing Hua University, Taiwan
5. Kim J, Moin P (1985) Application of a fractional-step method to incompressible navier-stokes equations. J Comput Phys 59(2):308–323
6. Hsu HW, Hwang FN, Wei ZH, Lai S-H, Lin CA (2011) A parallel multilevel preconditioned iterative pressure poisson solver for the large-eddy simulation of turbulent flow inside a duct. Comput Fluids 45(1):138–146
7. Pearson CE (1964) A computational method for time-dependant two-dimensional incompressible viscous flow problems. Report No. SRRC-RR-64-17, Sperry Rand Research Center, Sudbury, Massachusetts
8. Gavrilakis S (1992) Numerical simulation of low-Reynolds-number turbulent flow through a straight square duct. J Fluid Mech 244:101–129
9. Hsu HW, Hsu JB, Lo W, Lin CA (2012) Large eddy simulations of turbulent Couette-Poiseuille and Couette flows inside a square duct. J Fluid Mech 702:89101

Chapter 5
Turbulent Pressure-Driven Flow in a Square Duct at Low Reynolds Numbers

5.1 Introduction

Previous experimental studies on turbulent square duct flow have focused mainly on high Reynolds numbers for which a turbulence-induced eight-vortex secondary flow pattern exists in the cross-sectional plane. More recently, direct numerical simulations (DNS) have revealed that the flow field at Reynolds numbers close to transition can be very different; the flow in this 'marginally turbulent' regime may alternate between two states characterised by four vortices. This phenomenon has not previously been observed in experiments. It is the aim of this chapter, therefore, to provide experimental data on turbulent flow in a square duct at relatively low Reynolds numbers in order to investigate these marginally turbulent states and provide validation data for DNS. First, the onset criteria for transition to turbulence in square ducts is determined. In so doing, the potential importance of Coriolis effects on this process for low-Ekman number flows is highlighted. Data on the mean flow properties and turbulence statistics in both marginally and fully turbulent flow are then presented. The measurements have been obtained using laser Doppler velocimetry.

5.2 Transition to Turbulence

It is well known that a turbulent flow field can be split into a mean and fluctuating component. Steady laminar flow is characterised by no fluctuations, hence an increase in intensity of the fluctuating velocity component is a good measure for detecting the onset of transition to turbulence in ducts. Figure 5.1a and b show the variation of mean streamwise velocities, $\bar{u}$, normalised by the bulk (average) velocity, U_b, and the root mean square fluctuations, u_{rms}, with Reynolds number at the duct centreline ($y/h = 1.0$). The data have been obtained using water.

© Springer Nature Switzerland AG 2019
B. Owolabi, *Characterisation of Turbulent Duct Flows*,
Springer Theses, https://doi.org/10.1007/978-3-030-19745-2_5

Fig. 5.1 Onset criteria for square duct turbulent flow ($Ek \approx 1$, $y/h = 1$). **a** Variation of $\bar{u}/U_b$ with Reynolds number. **b** Variation of $u_{rms}/\bar{u}$ with Reynolds number. —— laminar flow analytical solution at $y/h = 1$; $\diamondsuit$ DNS of Gavrilakis [1]; —·—· numerical simulation of laminar flow at $Ek = 1$

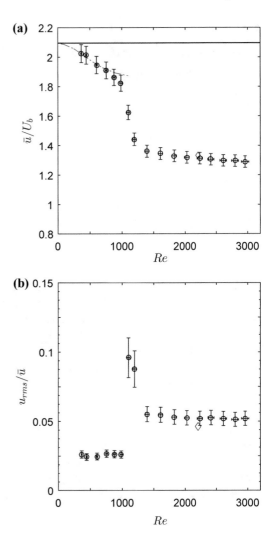

In Fig. 5.1a, a drop off in the values of $\bar{u}/U_b$ from the laminar flow analytical solution ([2], p. 113) (see Eq. 5.1), can be observed for $Re < 1000$.

$$u = U_b \frac{48}{\pi^3} \frac{\displaystyle\sum_{i=1,3,5,\ldots}^{\infty} (-1)^{(i-1)/2} \left[1 - \frac{\cosh(i\pi y/2a)}{\cosh(i\pi b/2a)} \right] \frac{\cos(i\pi z/2a)}{i^3}}{1 - \frac{192a}{\pi^5 b} \displaystyle\sum_{i=1,3,5,\ldots}^{\infty} \frac{\tanh(i\pi b/2a)}{i^5}} \tag{5.1}$$

This deviation from the analytical solution is surprising as the velocity fluctuations remain very low (about 2%), indicating that the flow is laminar (see Fig. 5.1b). Ideally, the fluctuations in a laminar flow should be zero. However, it is difficult to control all possible disturbances (such as those due to heat exchange, flow instabilities and vibration among others) in a real experiment. The measurement noise of the LDV system also contributes to the observed fluctuations. The deviation of the observed $\bar{u}/U_b$ from the analytical solution can be ascribed to Coriolis effects due to the earth's rotation as has been observed previously in pipe flows by Draad and Nieuwstadt [3]. They showed that the effect of rotation on a fully-developed laminar flow can be estimated using the Ekman number, a ratio of viscous to Coriolis forces:

$$Ek = \frac{\nu}{2\Omega D^2 \sin \alpha}, \qquad (5.2)$$

where Ω is the angular velocity of the earth ($7.272 \times 10^{-5}\text{s}^{-1}$), D is the duct width (0.08 m), α is the angle between the duct axis and the earth's rotation axis ($\approx 69°$) and ν the kinematic viscosity of the fluid. In this study, $Ek \approx 1$ was obtained for water ($\nu \approx 10^{-6}$ m^2 s^{-1}), hence Coriolis effects cannot be ignored in the laminar regime for this fluid. The Coriolis force brings about a distortion in the laminar flow velocity profile, which should normally be parabolic and symmetrical about the wall bisector, by introducing acceleration components in the wall-normal directions.

Transition can be said to take place at $Re \approx 1050$ as evidenced by a significant drop in the value of $\bar{u}/U_b$ from those of laminar flow. At this point, turbulent bursts are introduced into the laminar flow, hence the large fluctuation levels recorded in Fig. 5.1b. As the Reynolds number is increased, $u_{rms}/\bar{u}$ settles down to about 5%, the turbulence having become self sustaining, and $\bar{u}/U_b$ gradually approaches the fully-turbulent value obtained from the DNS of Gavrilakis [1]. In turbulent flow, inertial forces dominate and Coriolis effect becomes negligible.

To reduce the effect of Coriolis force in the laminar regime, a more viscous liquid—50% glycerol/water solution ($\nu \approx 6 \times 10^{-6}$ m^2 s^{-1})—was used resulting in $Ek \approx 7$. This fluid was employed for all subsequent experiments. Figure 5.2a and b show the variation of $\bar{u}/U_b$ and $u_{rms}/\bar{u}$ with Reynolds number. The experimental data for $\bar{u}/U_b$ at $y/h = 1$ can be observed to agree well with the laminar flow analytical solution until $Re \approx 800$ where there is a drop off once again due to the Coriolis effect. This is in agreement with the findings of Escudier et al. [4] who obtained laminar velocity profiles in pipe flow which were slightly asymmetric for Re as low as 540 for $Ek \approx 5$. Transition to turbulence can be observed to take place at $Re \approx 1250$ as indicated by an increase in velocity fluctuation levels as well as a significant drop in $\bar{u}/U_b$ from the laminar flow analytical solution. The difference in transition Reynolds number between water and 50% glycerol/water solution can be attributed to their dissimilar laminar base profiles prior to transition.

To determine the lowest Reynolds number for the onset of sustained turbulence, a trip rod, 10mm in diameter was introduced upstream (about 238h from the measurement section). The rod was placed vertically along the bisector of the horizontal walls, extending across the entire height of the duct. In this instance transition can be observed to take place at $Re \approx 940$. This value of Re is within the range given by Biau and Bottaro [5] and Uhlmann et al. [6]: $Re = 865$ and 1077 respectively.

Fig. 5.2 Onset criteria for square duct turbulent flow ($Ek \approx 7$). **a** Variation of $\bar{u}/U_b$ with Reynolds number. **b** Variation of $u_{rms}/\bar{u}$ with Reynolds number. Open symbols $y/h = 1$; closed symbols $y/h = 0.3$; □ trip rod introduced upstream; ○ no trip rod upstream; —— laminar flow analytical solution at $y/h = 1$; ······ laminar flow analytical solution at $y/h = 0.3$; ◇ DNS of Gavrilakis [1]; — · —· numerical simulation of laminar flow at $Ek = 7$

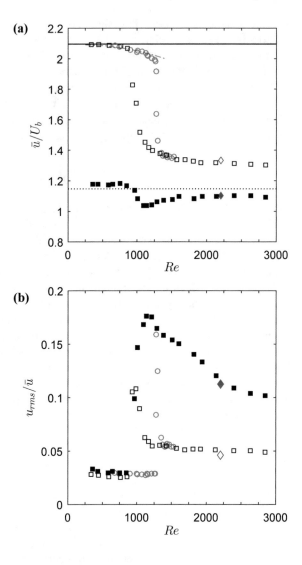

It is however higher than those where previous studies have shown the emergence of travelling wave solutions, which are known to be precursors to fully-developed turbulence.

At $y/h = 0.3$ (see Fig. 5.3), the streamwise velocity, normalized by the bulk velocity can be observed to be roughly constant in the laminar regime. It however drops off during transition before approaching the fully-turbulent value given by Gavrilakis [1] which, coincidentally, is close to the laminar flow value. The velocity fluctuations at $y/h = 0.3$ after transition are much higher than those at the duct centre and attain that of fully-turbulent flow more slowly (see Fig. 5.2b). This lends

Fig. 5.3 Cross section of
square duct showing
measurement location.
$y/h = 0.3$ is indicated by $\times$

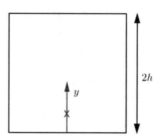

credence to the importance of the buffer layer in the transition dynamics of square
duct flow, as the largest turbulence intensities are observed in this region.

5.2.1 Numerical Simulation of Square Duct Laminar Flow Under the Influence of Coriolis Force

Following the approach of Draad and Nieuwstadt [3], a numerical simulation of fully
developed laminar square duct flow including the effect of the earth's rotation was
carried out in order to confirm the hypothesis that these effects cannot be neglected.
The governing Navier-Stokes equations for the flow are as follows:

$$\frac{Du_i}{Dt} = f_i - \frac{1}{\rho}\frac{\partial p}{\partial x_i} + \nu\frac{\partial^2 u_i}{\partial x_j^2}, \tag{5.3}$$

where u_i and f_i denote the velocity field and Coriolis force per unit mass respectively,
and p is the pressure term accounting for centrifugal forces. Considering the axis
system of Fig. 3.1c, and a unidirectional flow vector $[u(x, y, z), 0, 0]$, The Coriolis
force per unit mass is given by: $\vec{f} = -2\vec{\Omega} \times \vec{u}$, where $\vec{\Omega}$ is the earth's angular velocity
vector given by:

$$\vec{\Omega} = \Omega \begin{pmatrix} \cos\alpha_L \cos\alpha_N \\ \sin\alpha_L \\ -\cos\alpha_L \sin\alpha_N \end{pmatrix}, \tag{5.4}$$

α_L and α_N being the latitude (53° in our case) and the angle between the direction
of true north and the duct axis (100°) respectively. Hence:

$$\begin{pmatrix} f_x \\ f_y \\ f_z \end{pmatrix} = -2\Omega u(x, y, z) \begin{pmatrix} 0 \\ -\cos\alpha_L \sin\alpha_N \\ -\sin\alpha_L \end{pmatrix}. \tag{5.5}$$

The Navier-Stokes equations were solved using the commercial software, FLUENT,
for a domain: $D \times D \times 110D$ with $50 \times 50 \times 2160$ grid points (uniform in the cross-

sectional plane, but non-uniform in the streamwise direction such that the ratio of
the largest to the smallest cell is 7.5), using a pressure based coupled solver. This
software makes use of the finite volume approach. Spatial discretization of pressure,
gradient and momentum were carried out using the least square cell based, standard
and second order upwind schemes respectively.

Numerical simulation results at $y/h = 1$ for $Ek = 1$ and 7 are shown as dashed
lines in Figs. 5.1a and 5.2a respectively. The results show good agreement with the
experimental data. The deviation of $\bar{u}/U_b$ from the laminar flow analytical solution
can be observed to increase with Reynolds number. This is due to an increase in
laminar velocity profile asymmetry, the level of distortion being higher for $Ek = 1$.
Given the previous results of Draad and Nieuwstadt [3] and the excellent agreement
between experiment and simulation here for two different Ekman numbers, it can be
concluded that Coriolis forces can be significant in fully developed laminar square
duct flow with $D = 0.08$ m, especially for water. As a consequence, all the data
which follows is for 50% glycerol/water solution ($Ek \approx 7$).

5.3   Characteristics of Low Reynolds Number Turbulent Flows

Having determined the critical conditions for transition to turbulence, LDV measurements of the turbulence field at relatively low Reynolds numbers were obtained. At
each wall-normal location, data was collected for typically 30 min in the fully turbulent regime and up to 1 hour in the marginally turbulent state. The region below 8 mm
($y/h = 0.2$) from the duct wall was not accessible to the laser beams in the wall-
normal plane; hence, data on wall-normal velocity components in that area could not
be collected (see also [7]). Measurements of the mean and instantaneous velocity are
presented in Sect. 5.3.1 and Sect. 5.3.2 respectively, while turbulence intensity and
Reynolds stress measurements are shown in Sect. 5.3.3.

5.3.1   Mean Streamwise Velocity Measurements

Figure 5.4a shows the profile of streamwise velocity normalised by the bulk velocity,
along the wall bisector, for both marginally-turbulent ($Re = 1203$, $Re_\tau = 81$, where
Re_τ is the Reynolds number based on the mean friction velocity at the centre of a wall)
and fully-turbulent ($Re = 2230$, $Re_\tau = 161$) flow. The DNS results of Gavrilakis [1]
at a Reynolds number of 2205 ($Re_\tau = 162$) and Uhlmann et al. [6] at a Reynolds
number of 1205 ($Re_\tau = 84$) as well as the laminar flow analytical solution [2] are
also presented for the purpose of comparison. The flow was found to be symmetric
hence only the data from the lower half of the duct along the vertical bisector is
presented, y being the distance from the bottom wall as shown in Fig. 3.1c.

Fig. 5.4 Axial velocity profiles along the wall bisector. **a** In outer units. **b** In wall units. ∘ experiment at $Re = 1203$ ($Re_\tau = 81$); △ experiment at $Re = 2230$ ($Re_\tau = 161$); ...... DNS of Uhlmann et al. [6] at $Re = 1205$; —— DNS of Gavrilakis [1] at $Re = 2205$ ($Re_\tau = 162$); —— laminar flow analytical solution [2]. - - - - $u^+ = 2.5 ln y^+ + 5.5$; — · — · $u^+ = y^+$

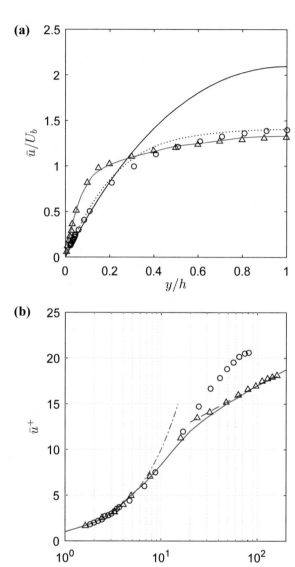

The experimental results show good agreement with DNS. It is interesting to note that the velocity gradient at the wall at $Re = 1203$ is very similar to that of laminar flow. However, beyond $y/h \approx 0.3$, the mean flow becomes markedly different from the laminar case. The values of $\bar{u}/U_b$ at the duct centre are 1.40 and 1.31 for marginally and fully-turbulent flow respectively.

In Fig. 5.4b, the same data is plotted in wall units, indicated by the superscript +, where the velocities have been normalised by local friction velocity (u_τ). Again, data at $Re_\tau = 161$ show excellent agreement with the DNS of Gavrilakis [1]. A large overshoot from the logarithmic law can be observed in the data for $Re_\tau = 81$, confirming that the wall shear stress distribution is characterised by a local minimum rather than maximum at the duct mid-point in marginally-turbulent flows, as shown by Pinelli et al. [8].

5.3.2 Instantaneous Velocity Measurements

The probability density functions (pdfs) of instantaneous streamwise velocity, u, normalised by either local mean, $\bar{u}$ or bulk velocity, U_b, for both marginally and fully-turbulent flow are presented in Fig. 5.5. At $Re = 1207$ (Fig. 5.5a), switching of the flow field between two states is confirmed by the pdf at $y/h = 0.3$ being bimodal. The point $y/h = 0.3$ has been chosen as it is where the largest difference between the two flow states can be observed in the DNS data of Uhlmann et al. [6]. The data were collected over a period of $12843h/U_b$ (3600 s), sufficiently long to capture many switches between states. The probability density function could be viewed as a combination of two pdfs, one for each flow state; but separating the two is not a trivial task as there is a significant overlap. It can be partially achieved by plotting a pdf of modal velocities (dotted line in Fig. 5.5a). The time series of velocity was split into small intervals of the order of $43h/U_b$ in length, and the modal velocity (i.e. the most frequently occurring velocity) in each interval was then computed. The resulting pdf shows the bimodal form created by the two flow states more clearly, although there is still overlap. The values of $u/\bar{u}$ at the peaks correspond reasonably well to those extracted from the time averages of Uhlmann et al. [6] as shown by the dashed vertical lines in Fig. 5.5a.

A bimodal pdf also occurs at $y/h = 0.2$ but beyond $y/h = 0.4$, this dual-peak feature fades away at higher distances from the wall as the two flow states become increasingly similar (see Fig. 5.5b). In contrast, the pdf for fully-turbulent flow at $Re = 2234$ and $y/h = 0.3$ (Fig. 5.5c) is unimodal, indicating that the flow exists in only one state. In this study, bimodal pdfs at $y/h = 0.3$ were observed between $Re = 1097$ and $Re = 1370$ as shown in Fig. 5.5c. The two peaks are mostly of different heights, indicating that the flow spends more time in one state than in the other, the largest peak gradually shifting from left to right as the Reynolds number is increased (see Fig. 5.5c).

The joint probability density function of streamwise and wall-normal velocities at $Re = 1290$ for data collected over a period of $7004h/U_b$ is shown in Fig. 5.6. A dual peak can be clearly seen, corresponding to two flow states, A and B. In state A, streamwise velocities lower than the long term mean (indicated as a dashed line) are highly probable to occur alongside wall-normal velocities which are higher than the long term mean (shown as a solid black line). In state B, streamwise velocities higher than the long term average have a higher probability of occurring together with essentially zero wall-normal velocities at this measurement location. This behaviour is consistent with the findings of Uhlmann et al. [6] with regard to the streak-vortex

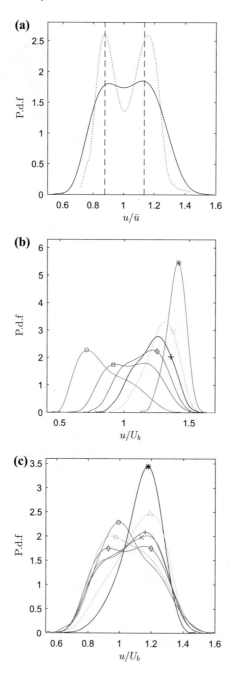

Fig. 5.5 Probability density functions of u/U_b. **a** $Re = 1207$ at $y/h = 0.3$. Dotted line is the pdf of modal velocities and dashed lines correspond to short time averages for each state from the DNS data of Uhlmann et al. [6] **b** $Re = 1203$ at $y/h = 0.2, 0.3, 0.4, 0.5, 0.6$ & 1. **c** $Re = 1097, 1125, 1207, 1290, 1370, 1596$ and 2234 and $y/h = 0.3$

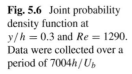

Fig. 5.6 Joint probability density function at $y/h = 0.3$ and $Re = 1290$. Data were collected over a period of $7004h/U_b$

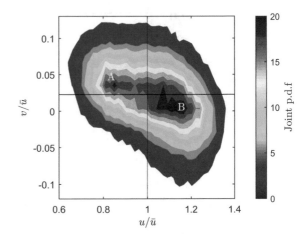

arrangement for any given pair of opposite walls during marginally-turbulent flow. The flow alternating between periods of high turbulence activity, characterised by a low velocity streak at the wall bisector flanked by vortices (corresponding to state A), and more quiescent periods where there are no streaks and the flow is essentially unidirectional at the measurement location (state B).

Variation of streamwise velocity skewness with distance from the duct wall is shown in Fig. 5.7a. The data for $Re = 2230$ show excellent agreement with the DNS of Gavrilakis [1], having positive values very close to the wall and becoming negative beyond $y/h \approx 0.075$. For $Re = 1203$, the velocities are more positively skewed in the near wall region up to $y/h \approx 0.7$, this is due to the presence of the two flow states at the marginally turbulent Reynolds number as shown by the pdfs in Fig. 5.5. Beyond $y/h \approx 0.7$, similar values of skewness are observed in both flows.

Streamwise velocity autocorrelation functions ($R_{u'u'}$) for both marginally and fully-turbulent flows at $y/h = 0.3$ are shown in Fig. 5.7b. The marginally-turbulent flow is correlated over a larger time period reinforcing the evidence that the flow remains in a particular state for a significant period of time. The integral time scales have been computed by integrating the autocorrelation functions up to the first zero crossing. The integral time scale for marginally-turbulent flow ($2.24h/U_b$) is larger than fully-turbulent flow ($0.89h/U_b$). This longer integral time is an approximate quantification of the longer "memory" in the marginally-turbulent flow state.

5.3.3 Turbulence Intensity and Reynolds Stress Measurements

The streamwise and wall-normal turbulence intensities at $Re_\tau = 81$ ($Re = 1203$) and $Re_\tau = 161$ ($Re = 2230$) as a function of distance from the duct wall along the bisector, in outer units, are shown in Fig. 5.8a. It can be observed that u_{rms}/U_b for

Fig. 5.7 a Axial velocity
skewness along the wall
bisector. ∘ experiment at
$Re = 1203$; △ experiment at
$Re = 2230$; —— DNS of
Gavrilakis [1] at $Re = 2205$.
b Streamwise velocity
autocorrelation at
$y/h = 0.3$: ——
$Re = 1207$; ——
$Re = 2234$; Δt is the time
lag

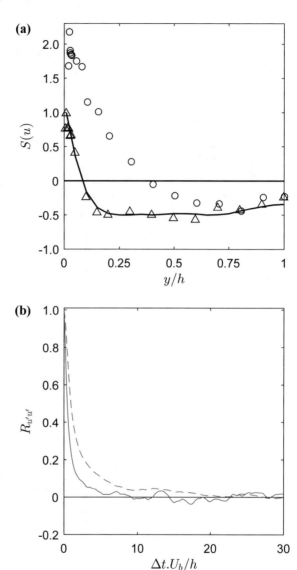

fully turbulent flow shows good agreement with the DNS of Gavrilakis [1], attaining
a peak at $y/h \approx 0.1$. This peak value is however slightly higher than the DNS result.
In the marginally-turbulent case, there appears to be a shift in the location of the
maximum value of u_{rms}/U_b away from the wall to $y/h \approx 0.3$. Close to the duct wall,
the streamwise velocity fluctuations are lower than in fully turbulent flow but become
larger as the peak value is approached and remains so across the remaining portion of
the duct, creating an impression that the turbulence level is higher. Figure 5.8b shows
the same data plotted as a function of distance along the bisector in wall units. In this

Fig. 5.8 Turbulence intensities along the wall bisector. **a** as a function of y/h, **b** as a function of y^+. ○ experiment at $Re_\tau = 81$; △ experiment at $Re_\tau = 161$; —— DNS of Gavrilakis [1] at $Re_\tau = 162$. Open and closed symbols represent streamwise and wall-normal turbulence intensities respectively

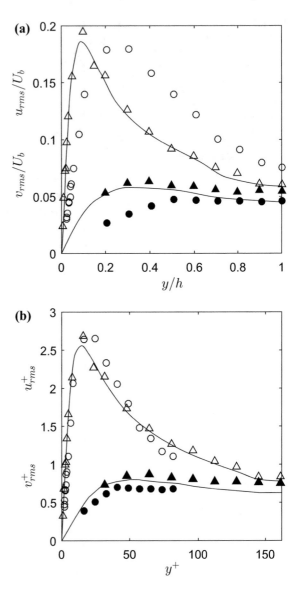

case, the streamwise turbulence intensities, u^+_{rms} are very similar in both flows, with u^+_{rms} at $Re_\tau = 81$ becoming lower than the fully turbulent values beyond y^+ of 63.

The wall-normal turbulence intensities (v_{rms}) which are significantly smaller, are almost constant across the duct and no distinct maxima can be identified. The wall-normal turbulence intensities at $Re_\tau = 161$ show good agreement with DNS data, they are however a bit higher than the simulation values towards the duct centre and larger than those at $Re_\tau = 81$. The results indicate that near the centre of the duct,

Fig. 5.9 Variation of Reynolds shear stress along the wall bisector: o experiment at $Re = 1203$; $\triangle$ experiment at $Re = 2230$; —— DNS of Gavrilakis [1] at $Re = 2205$

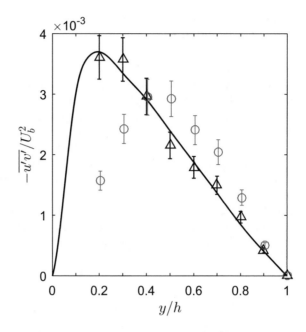

the turbulence is nearly isotropic, as shown by the closeness of u_{rms}^+ and v_{rms}^+ values at high y^+. The level of isotropy becomes higher with increasing Reynolds number.

Figure 5.9 shows the variation of Reynolds shear stress along the wall bisector. The data have been normalised by the square of the bulk velocity. Since only the value of local friction velocity (u_τ) at the wall mid-point is available in this experiment, normalising by u_τ^2 will not provide a true picture of Reynolds shear stress distribution. The data for $Re = 2230$ is in good agreement with DNS. At $Re = 1203$, Reynolds shear stress can be observed to be lower for $y/h < 0.4$. There is also a shift in location of the maximum away from the duct wall. As the duct centre is approached, Reynolds shear stress drops to zero for both flows as required by symmetry.

5.4 Summary

The behaviour of turbulent flow in a square duct at relatively low Reynolds numbers has been studied. The results for both marginally-turbulent flow at $Re_\tau = 81$ and fully-turbulent flow at $Re_\tau = 161$ show good agreement with the DNS data of Uhlmann et al. [6] and Gavrilakis [1] respectively. It has been shown that the onset of turbulence can be significantly affected by Coriolis effects due to the earth's rotation. This is as a result of differences in the laminar base flow velocity profiles at different values of Ekman number prior to transition. A limiting Reynolds number of about

940 for transition was observed at $Ek \approx 7$. This value is within the range obtained in previous numerical studies.

In marginally-turbulent flow a mean flow very similar to laminar flow has been observed in the vicinity of the duct wall, as well as a large overshoot from the logarithmic law nearer the centre. Switching between two flow states, originally predicted by the DNS of Uhlmann et al. [6], is confirmed by bimodal pdfs of streamwise velocity at certain distances from the duct wall and a joint pdf of streamwise and wall-normal velocity which features two peaks corresponding to each of the two states: one essentially unidirectional ($v^+ \approx 0$) at the measurement location and the other containing a significant secondary flow component ($v/\bar{u} \approx 0.03$).

It has been shown that marginally turbulent flow is more correlated than a fully turbulent flow as indicated by its longer integral time scale. Similar levels of u^+_{rms} have been observed in both flows at different y^+. The Reynolds stresses in the near wall region are however lower in the former but become very similar to those of fully turbulent flow as the duct centre is approached.

References

1. Gavrilakis S (1992) Numerical simulation of low-Reynolds-number turbulent flow through a straight square duct. J Fluid Mech 244:101–129
2. White FM (2006) Viscous fluid flow, 3rd edn. McGraw-Hill, New York
3. Draad A, Nieuwstadt F (1998) The earth's rotation and laminar pipe flow. J Fluid Mech 361:297–308
4. Escudier M, Poole R, Presti F, Dales C, Nouar C, Desaubry C, Graham L, Pullum L (2005) Observations of asymmetrical flow behaviour in transitional pipe flow of yield-stress and other shear-thinning liquids. J Non-Newt Fluid Mech 127(2):143–155
5. Biau D, Bottaro A (2009) An optimal path to transition in a duct. Phil Trans R Soc Lond A 367(1888):529–544
6. Uhlmann M, Pinelli A, Kawahara G, Sekimoto A (2007) Marginally turbulent flow in a square duct. J Fluid Mech 588:153–162
7. Escudier M, Smith S (2001) Fully developed turbulent flow of non-Newtonian liquids through a square duct. Proc R Soc A 457(2008):911–936
8. Pinelli A, Uhlmann M, Sekimoto A, Kawahara G (2010) Reynolds number dependence of mean flow structure in square duct turbulence. J Fluid Mech 644:107–122

Chapter 6
Turbulent Wall-Driven Flows

6.1 Introduction

While there have been several studies on pressure-driven (Poiseuille) flow in a square duct, the wall-driven (Couette) case remains largely unexplored. Large eddy simulations of Couette-Poiseuille flow by Hsu et al. [1] and Lo and Lin [2] revealed significant changes to the secondary flow in the duct, near a moving wall; the Reynolds numbers considered were, however, quite high ($Re > 4000$), hence the flow characteristics close to transition were not documented. As shown in Chap. 5, "marginally-turbulent" Poiseuille flow in a square duct is characterised by a switching of the flow field between two states, each having a secondary flow pattern which is very different from what is observed at high Reynolds numbers. An interesting question is whether this switching also exists in wall-driven flow.

In this chapter, the results of direct numerical simulations of purely wall-driven turbulent flows in a square duct are presented. First, the critical conditions for self-sustained turbulence are determined. Motivated by the findings on the existence of bi-stable states in "marginally-turbulent" Poiseuille square duct flows, the turbulence field is then characterised at relatively low Reynolds numbers with a view to determining whether this phenomenon is a ubiquitous feature of wall-bounded flows with no spanwise homogeneity. For a mean secondary flow to exist, a duct of finite aspect ratio is required. In this regard, a square duct is arguably the simplest and most widely studied geometry, thus it was selected to allow for direct comparison with results on pressure-driven flow [3]. The focus is on the case in which a pair of opposite counter-moving walls translating with the same speed (U_w) drive the flow. This configuration results in a zero net transport of fluid through the duct.

Figure 6.1 shows the problem setup. The streamwise direction is x while y and z are the transverse and spanwise directions, respectively, with u, v and w being the corresponding velocity components. The duct has a dimension of $2h \times 2h \times L_x$, where L_x is the streamwise extent. A Reynolds number, Re_w, given by $Re_w = U_w h/\nu$ can be defined, where U_w is the velocity of a moving wall.

© Springer Nature Switzerland AG 2019
B. Owolabi, *Characterisation of Turbulent Duct Flows*,
Springer Theses, https://doi.org/10.1007/978-3-030-19745-2_6

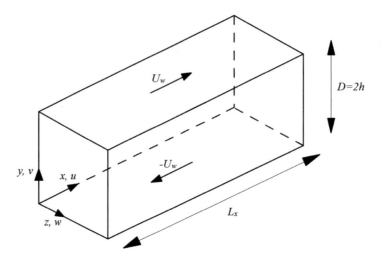

Fig. 6.1 Coordinate system and geometry

6.2   Critical Conditions for Self-sustained Turbulence

In this section, the results of simulations carried out to determine the critical Reynolds number (Re_c) for transition to turbulence and to obtain an estimate of the typical length scales (L_c) of the smallest structures required for the sustenance of a turbulent state in wall-driven square duct flow are presented. The value of L_c places a lower bound on the length of the box that can be used for a turbulent flow simulation.

Given the sub-critical nature of the transition, initial conditions are very important. Following the approach of [3–5], the simulations were initiated using, as starting conditions, the fully developed turbulent flow field at a high Reynolds number and/or duct length (L_x). The Reynolds number was then gradually varied in successive runs while keeping L_x constant until the flow re-laminarised. In other cases, L_x was varied while holding Re_w constant. The fluctuations in the streamwise and wall-normal velocity at different locations in the duct were monitored; upon re-laminarisation, these dropped to zero. The time steps employed in each simulation were such that the Courant-Friedrichs-Lewy (CFL) number was less than 0.3, and the flow was allowed to evolve for a time of at least 3500 h/U_w. Similar integration times have been employed by Uhlmann et al. [3] and Jiménez and Moin [4] in their studies on turbulence sustenance in minimal flow units.

Figure 6.2 shows the flow states for different combinations of Re_w and L_x. The data points at the boundary between laminar and turbulent states are connected with dashed lines. For turbulent states away from this boundary, a mesh with 96 × 96 cross-sectional divisions was found to be sufficient to obtain a sustained turbulent state. At $Re_w = 1500$ and $L_x = 4 \pi h$, the turbulence statistics obtained using this grid were compared with that from a grid having 128 × 128 cross-sectional divisions

Fig. 6.2 Critical conditions
for turbulence. Filled
symbols represent laminar
states while unfilled ones are
the turbulent states. $\triangle$,
$N_z = N_y = 128$ grid cells;
$\bigcirc$, $N_z = N_y = 96$ grid
cells;
- - - -, laminar/turbulent
boundary

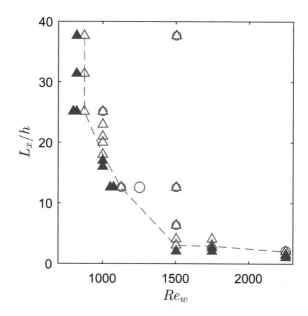

and there was good agreement between the two. In both cases, the number of cells
in the streamwise direction (N_x) was 160.

From Fig. 6.2, it can be observed that the lowest Reynolds number (Re_c) at which
a turbulent state can be sustained is about 875. There is, however, an uncertainty
of about ± 25 (since the Reynolds number was varied in steps of 50). For values of
Re_w less than the critical, a re-laminarisation of the flow occurred irrespective of the
domain length. Given that the same estimate of Re_c was obtained in boxes of length
$8\,\pi h$, $10\,\pi h$ and $12\,\pi h$, its value is not expected to change much in a longer box.
To further check that Re_c is independent of the grid, the simulation at $Re_w = 825$
and $L_x = 10\,\pi h$ (the laminar data point just before Re_c) was repeated using two
different meshes having 128×128 and 256×256 cells in the cross-sectional plane
and 320 cells in the streamwise direction and a re-laminarisation of the flow was
observed in both. The grid independence of L_c at the Reynolds numbers considered,
was also confirmed by repeating the cases where re-laminarisation occured, with a
mesh having 128×128 cross-sectional divisions. Table 6.1 gives the parameters for
the eight data points connected by the dashed line in Fig. 6.2.

The transition Reynolds number is much greater than that in plane Couette flow
($Re_c \approx 325 - 370$, see [6–8]), thus showing the importance of side walls in stabilising the flow. For large aspect ratio ducts, where end effects are negligible, Re_c is
expected to be similar to that in plane Couette flow. To verify this, simulations were
carried out in two ducts of aspect ratios 2:1 and 4:1 and lengths $4\,\pi h$ and a turbulent
state was observed to be maintained at Re_w as low as 625 and 500, respectively. For
the pressure-driven case, [9] suggested that a duct with an aspect ratio of at least 24

Table 6.1 Simulation parameters for data points on the laminar/turbulent boundary. L_x is the length of computational domain Δy^+ is the grid resolution along the bisector of the moving wall and Δx^+, is that in the streamwise direction. The superscript, + refers to normalisation by wall units defined in terms of the local friction velocity at the centre of a moving wall. N_x, N_y and N_z are the number of grid points in the x, y and z directions, respectively. Re_τ is defined in terms of the average value of the shear stress at the centre plane of the moving walls

Re_w	L_x	$N_y \times N_z \times N_x$	Re_τ	Δy^+	Δx^+
875	$12\,\pi h$	$128 \times 128 \times 352$	52	$0.04 - 1.91$	5.80
875	$10\,\pi h$	$128 \times 128 \times 320$	52	$0.04 - 1.91$	5.16
875	$8\,\pi h$	$128 \times 128 \times 320$	53	$0.04 - 1.92$	4.17
1000	$18\,h$	$128 \times 128 \times 192$	59	$0.04 - 2.16$	5.60
1125	$4\,\pi h$	$96 \times 96 \times 160$	65	$0.07 - 2.50$	5.19
1125	$4\,\pi h$	$128 \times 128 \times 160$	65	$0.05 - 2.38$	5.19
1500	$3\,h$	$128 \times 128 \times 96$	88	$0.07 - 3.21$	2.81
1750	$3\,h$	$128 \times 128 \times 64$	101	$0.08 - 3.69$	4.89
2250	$2\,h$	$96 \times 96 \times 64$	125	$0.13 - 4.65$	4.52
2250	$2\,h$	$128 \times 128 \times 64$	125	$0.10 - 4.57$	4.04

is required to obtain results comparable to those in a channel; thus it is expect that a similar aspect ratio will be required in Couette flow.

Close to transition, L_c can be observed to be larger than at higher Reynolds numbers, indicating an increase in the length scales of the turbulence structures, hence longer computational boxes are needed for the simulations. For $Re_w > 1500$, L_c is about $3\,h$. The flow in such minimal units are not realisable in the laboratory; rather, they are the basic building blocks of wall-bounded turbulent flows [4]. To simulate a physically realisable turbulent flow, the domain length would have to be much longer than L_c such that a decorrelation of the turbulence statistics is achieved.

6.3 Turbulent Flows at Low Reynolds Numbers

Having determined the critical conditions for self-sustained turbulence, the turbulent flows at relatively low Reynolds numbers are investigated. In the following sections, the results obtained at $Re_w = 1500$ in a computational domain of length $12\,\pi h$ are

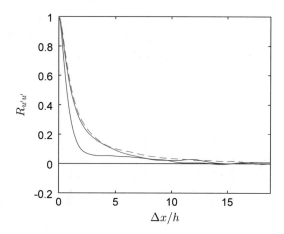

Fig. 6.3 Two-point autocorrelation coefficients of streamwise velocity fluctuations ($R_{u'u'}$) at $Re_w = 1500$ in a domain of length 12 πh. ____, $y/h = 0.2$; _ _ _ _, $y/h = 1.8$; ____, $y/h = 1$. $R_{u'u'}$ has been computed at points along the moving-wall bisector. Only half of the computational domain is shown

presented. This box was found to be long enough to allow for the decorrelation of velocity fluctuations (see Fig. 6.3). A mesh having $96 \times 96 \times 352$ cells has also been used. For this grid it has been verified that the cell sizes are of the order of the Kolmogorov length scale (η) given by $\eta = (\nu^3/\epsilon)^{1/4}$, where ϵ is the rate of turbulent kinetic energy dissipation. At the wall, ϵ was estimated from the viscous diffusion term in the turbulent kinetic energy equation [10] but away from the wall, ϵ was obtained by assuming turbulent kinetic energy production to be equal to dissipation. From this analysis, the grid resolution ranged from about $0.16\,\eta$ to $1.8\,\eta$. Along the bisector of the moving walls, Δy^+ varied from about 0.09 at the wall to 4.05 at the centre of the duct, while in the streamwise direction, Δx^+ was about 8.91.

6.3.1 Secondary Flow Pattern

Figure 6.4 shows the mean velocity fields, the velocity vectors indicating the secondary flow pattern. An alternation of the flow field in time between two states can be observed, one being a mirror reflection of the other, and the flow remaining approximately symmetrical about the common bisector of the moving walls (see Fig. 6.4a and b). The velocity fields have been obtained by averaging over two separate intervals of lengths 1050 h/U_w and 1506 h/U_w, respectively, during which the flow was continuously in each state. Hereafter the states corresponding to Fig. 6.4a is referred to as A and that shown in Fig. 6.4b as B. In either case, a pair of large counter-rotating vortices associated with a moving wall dominates the entire flow field, transporting momentum from the wall to the interior of the duct, and another pair of smaller vortices is located at the opposite wall. The time spent in each state can be rather long. As the flow switches states, the large vortices shrink and are pushed towards the corners, while maintaining their rotation sense and the vortex pair at the opposite wall

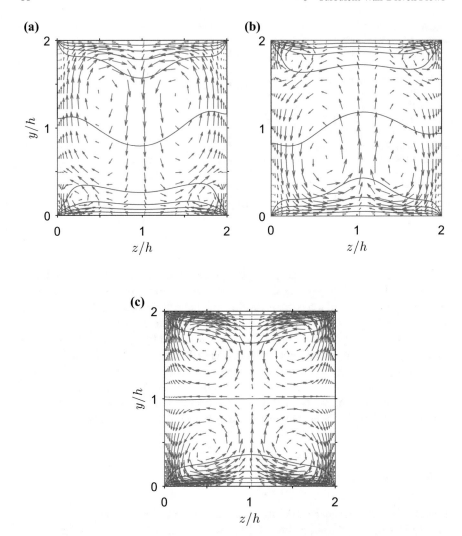

Fig. 6.4 Contours of mean streamwise velocity normalised by the wall velocity, U_w, and secondary flow vectors at $Re_w = 1500$ and $L_x/h = 12\,\pi$: **a** averaging interval 1050 h/U_w, **b** average over a different interval of length 1506 h/U_w (flow is in a different state), **c** averaging interval 12960 h/U_w (including both previous intervals). Contours range from -1 at the lower wall to 1 at the upper wall, with increment 0.2. For clarity, vectors are shown at every third grid point

become enlarged. Averaging over long times results in a four-vortex secondary flow pattern, symmetrical about the wall bisectors (see Fig. 6.4c, where the integration time is 12960 h/U_w).

To quantitatively identify the two states, the magnitude $\mathcal{S}(t)$ of the secondary flow in the central part of the duct is defined, with a sign assigned, depending on the value of the wall-normal velocity component:

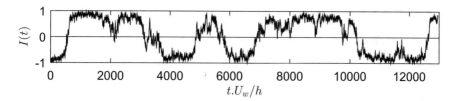

Fig. 6.5 Variation of the indicator function, $I(t)$ with non-dimensional time at $Re_w = 1500$ and $L_x/h = 12\pi$

$$S(t) = \begin{cases} \sqrt{\tilde{v}^2 + \tilde{w}^2}, & \tilde{v} > 0 \\ -\sqrt{\tilde{v}^2 + \tilde{w}^2}, & \tilde{v} < 0, \end{cases} \tag{6.1}$$

where the tilde symbol represents instantaneous spatial averaging in the streamwise direction. An indicator function given by Eq. 6.2 is introduced:

$$I(t) = \frac{\iint_R S(t) dy dz}{\iint_R |S(t)| dy dz}, \tag{6.2}$$

where R is the region bounded by the lines $z/h = 0.6$, $z/h = 1.4$ and the upper and lower walls, where the secondary flow is mostly in the direction normal to the moving walls (see Fig. 6.4a and b). $I(t)$ ranges between -1 and 1, negative values corresponding to state A and positive values, B. Based on the above criteria, conditional averaging was carried out to separate the two states. Figure 6.5 shows the evolution of the indicator function over the interval for which the flow fields in Fig. 6.4 have been computed. The average values of I in Fig. 6.4a and b are -0.775 and 0.709, respectively. The change in state can be observed to occur in an irregular fashion; however, the time spent in each state is of the order of $1000\ h/U_w$.

Statistical convergence of the long-time-averaged flow fields was checked by computing the velocity magnitude and turbulent kinetic energy (t.k.e) from data accumulated over different times (see Fig. 6.6, where data along the moving-wall bisector is shown). The simulation was run for a period of up to $38650\ h/U_w$. For times greater than $28140\ h/U_w$, the differences in the computed statistics were found to be less than 1%, hence the analyses in the following sections are based on flow fields accumulated over integration periods larger than $28140\ h/U_w$.[1]

6.3.2 Origin of the Secondary Flow

Associated with the secondary flows in non-circular ducts is a mean streamwise component of vorticity; hence by examining the vorticity transport equation, an

---

[1]For other values of Re_w, a similar check would have to be carried out to ensure convergence of the long-time statistics.

Fig. 6.6 Convergence test: **a** mean velocities computed from different integration times; **b** turbulent kinetic energy computed from different integration times. Data along the moving-wall bisector is presented. ×, $y/h = 1.0$; □, $y/h = 1.2$; *, $y/h = 1.4$; o, $y/h = 1.6$; +, $y/h = 1.8$

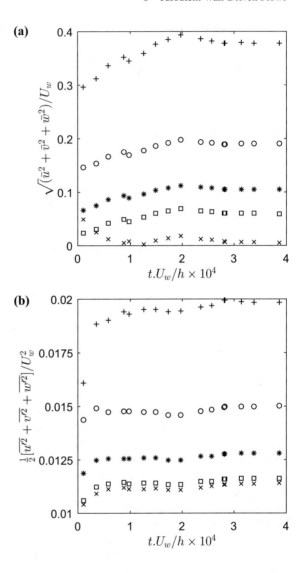

insight into the origin of these motions can be obtained. For fully developed flow in a straight duct, the equation reads

$$\underbrace{\bar{v}\frac{\partial \bar{\Omega}_x}{\partial y} + \bar{w}\frac{\partial \bar{\Omega}_x}{\partial z}}_{C} - \underbrace{\nu\left(\frac{\partial^2}{\partial y^2} + \frac{\partial^2}{\partial z^2}\right)\bar{\Omega}_x}_{D} + \underbrace{\left(\frac{\partial^2}{\partial y^2} - \frac{\partial^2}{\partial z^2}\right)\overline{v'w'}}_{P_1} + \underbrace{\frac{\partial^2}{\partial y \partial z}(\overline{w'^2} - \overline{v'^2})}_{P_2} = 0,$$

(6.3)

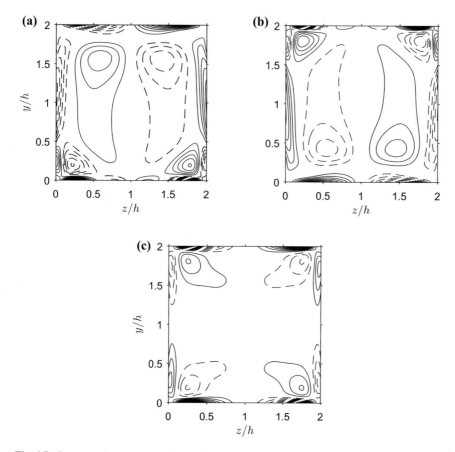

Fig. 6.7 Contours of mean streamwise vorticity at $Re_w = 1500$. **a** and **b** correspond to states A and B, respectively, while (**c**) is the contour plot for the long-time-averaged flow field. The data have been normalised by the maximum absolute vorticity; contours range from -1 to 1 with increment 0.1; negative values dashed. Length of computational domain is $12\,\pi h$

where $\bar{\Omega}_x = \frac{\partial \overline{w}}{\partial y} - \frac{\partial \overline{v}}{\partial z}$ is the mean streamwise vorticity, and the prime symbol as well as the overbars represent fluctuating velocity components and time averaging, respectively. The first two terms, C, on the left hand side of Eq. (6.3) represent the convection of streamwise vorticity by the secondary motion itself. Together with the viscous diffusion term, D, these quantities are mainly involved in the redistribution of vorticity within the duct. P_1 and P_2 represent the contribution of the Reynolds cross-stream shear stress and the anisotropy of the cross-stream normal stresses, respectively. They act to either produce or destroy streamwise vorticity.

Figure 6.7 shows the contours of $\bar{\Omega}_x$. It can be observed that they closely match the secondary flow patterns shown in Fig. 6.4, except close to the walls where there is an inversion of the vorticity sign. The corner vortices are associated with higher values of vorticity, but the maximum is located by the moving wall. Figure 6.8a and

Fig. 6.8 Contours of production terms in the mean streamwise vorticity transport equation, normalised by $(U_w/h)^2$. **a** and **b** show the normal (P_2) and shear stress (P_1) terms respectively, while the flow is in state A. **c** and **d** show P_2 and P_1 respectively, for the long-time-averaged flow field. Contours range from -0.1 to 0.1, with increment 0.02; negative values are shown by the dashed curve

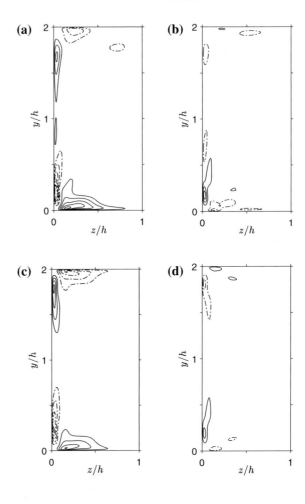

b show the distribution of the normal and shear-stress terms of the vorticity transport equation, respectively, while the flow is in state A (note that the corresponding plots for state B can be obtained by a π radians rotation about the duct's axis). The plots for the long-time-averaged data are shown in Fig. 6.8c and d. Since the flow is symmetrical about the common bisector of the moving walls, only half of the computational domain is shown. In state A, close to the lower corner, where the small vortices are located and across most of the lower wall, the contribution of P_2 to the production of streamwise vorticity can be observed to be larger than that of P_1. Its maximum value, which is about 1.8 times larger than the maximum P_1 occurs at $z/h = 0.023$, $y/h = 0.190$. However, at the upper wall, where the large vortices are located, P_1, though relatively small in magnitude, accounts for the vorticity production away from the corner. Switching between states results, in the long run, in a flow field in which streamwise vorticity production is dominated by gradients of the anisotropy of the Reynolds normal stresses (see Fig. 6.8c and d).

Fig. 6.9 Budget of terms in the mean streamwise vorticity transport equation along lines parallel to the side wall and passing through the centre of the vortices: **a** at $z/h = 0.25$, while the flow is in state A, **b** at $z/h = 0.6$ [the same state as (**a**)], $+$, normal stress (P_2); ———, shear stress (P_1); $\times$, viscous diffusion; -.-.-, convection; - - - -, balance; ...., $\bar{\Omega}_x/(6U_w/h)$, **c** at $z/h = 0.4$ in the long-time-averaged flow field. Legend as in (**a**) and (**b**)

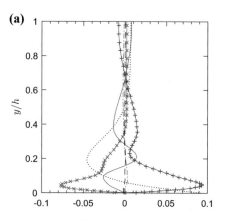

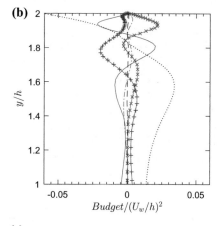

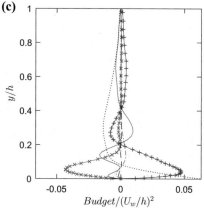

The contributions of terms in Eq. (6.3) are quantitatively shown in Fig. 6.9. Plots of the streamwise vorticity are also included for reference. Consider a line parallel to the y axis, located at $z/h = 0.25$, and passing through the centre of the corner vortex. Close to the lower wall, along this line, the production of vorticity is dominated by the anisotropy term, P_2 (see Fig. 6.9a), the maximum value occurring at $y/h \approx 0.05$ where there is a change in the vorticity sign. P_2 is balanced by viscous diffusion and the shear-stress term which is much smaller in magnitude. It can be observed that the contributions of P_2 and P_1 are of opposite signs across most of the duct, one acting as a source term and the other, having the same sign as the viscous diffusion term, acting as a sink. Beyond $y/h \approx 0.36$, P_1 and P_2 are roughly of the same magnitude. The convection term is very small across the entire length shown, and so is D beyond $y/h \approx 0.36$. Figure 6.9b shows the terms in the vorticity transport equation along a vertical line, located at $z/h = 0.6$, passing through the centre of the large vortex associated with the upper wall. Close to the wall, P_1 can be observed to account for most of the vorticity production as earlier discussed, its maximum value occurring at $y/h \approx 1.94$. Again, balance is maintained by the viscous diffusion term, while the convection term is very small. Below $y/h \approx 1.94$, P_2 increases, attaining a maximum at $y/h \approx 1.79$, beyond which it is roughly of the same magnitude as P1. For the long-time-averaged flow field along a line passing through the centre of a secondary flow vortex at $z/h = 0.4$ (see Fig. 6.9c), a trend similar to that of Fig. 6.9a can be observed. From the foregoing, it can be concluded that gradients of the Reynolds secondary shear stress, $\overline{v'w'}$ away from the corners play a key role in the formation of the large secondary flow vortices. On the other hand, the corner vortices emerge as a result of gradients of $\overline{w'^2} - \overline{v'^2}$.

6.3.3 Flow Statistics and Turbulence Structure

Streamwise velocity profiles along the moving-wall bisector, normalised by local friction velocity are presented in Fig. 6.10a. The plane Couette flow (channel) data of Uhlmann et al. [11] and Jiménez and Moin [12] at $Re_\tau = 125$ and 171, respectively and Poiseuille flow data of Chap. 5 at $Re_\tau = 81$ are also shown for the purpose of comparison. In state A, at the lower half of the duct, the velocity profile can be observed to be in good agreement with those in the channel. The data matches the logarithmic scaling law: $|\bar{u} - U_w|^+ = 2.44 \, ln(y^+) + 5.1$, in the region: $30 < y^+ < 69$. At the duct's upper half (equivalent to the lower half in state B), however, an overshoot from the log law can be observed. As will be shown later, the local shear stress at the upper moving wall is smaller than at the lower one. Therefore, normalising by the local friction velocity results in higher non-dimensional velocities being obtained. Furthermore, the Reynolds number based on the local friction velocity ($Re_\tau = 70$) is smaller than at the lower wall ($Re_\tau = 90$) hence this deviation from the classical profile is a low Reynolds number effect. For the long-time-averaged flow field, the velocity profile lies between those of the bi-stable states.

Fig. 6.10 Turbulence statistics: **a** Streamwise velocity profiles, **b** Root-mean-square velocity fluctuations. ——, profile from the lower half of the duct while the flow is in state A; – – – –, profile from upper half of the duct while the flow is in state A; -.-.-.-, long-time-averaged flow field; $\square$, Plane Couette flow DNS of Avsarkisov et al. [11] at $Re_\tau = 125$; $\star$, plane Couette flow DNS of [12] at $Re_\tau = 171$; $\triangle$, long-time-averaged marginally turbulent square duct Poiseuille flow experiment of Chap. 5; -.-.-.-, $|\bar{u} - U_w|^+ = y^+$; ...., $|\bar{u} - U_w|^+ = 2.44.ln(y^+) + 5.1$

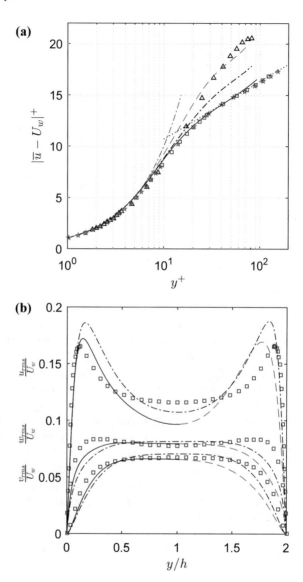

Root mean square velocity fluctuations along the moving-wall bisector normalised by the wall velocity are shown in Fig. 6.10b. Larger streamwise fluctuations can be observed across most of the duct in the long-time-averaged flow field. These are induced by the intermittency in the flow. In one state, the velocities fluctuate about a mean which is higher than the long-term one, while in the other, the fluctuations are about a mean value lower than the long-time-average. Hence in the long run, the deviations from the long-term mean are large (see Fig. 6.11, where the probability density functions of instantaneous streamwise velocity at $y/h = 0.4$ are shown). In

Fig. 6.11 Probaility density function of instantaneous streamwise velocity, $\tilde{u}$, at $y/h = 0.4$. Legend as in Fig. 6.10. Arrows indicate fluctuations about the short and long-time averages

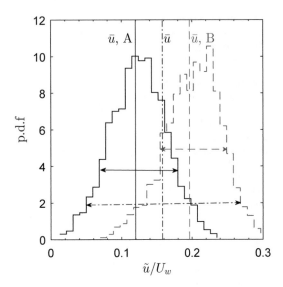

state A, the peaks in u_{rms} ($u_{\mathrm{rms}}/U_w \approx 0.172$ and 0.170 near the lower and upper walls, respectively) can be observed to be slightly higher than the ones in the plane Couette flow data of Avsarkisov et al. [11] ($u_{\mathrm{rms}}/U_w \approx 0.165$). The local maximum in the u_{rms}/U_w profile exists closer to the lower wall than the upper wall. As earlier shown, the Reynolds numbers based on the local friction velocity is larger at the bottom, hence this trend is consistent with findings in wall-bounded turbulent flows which show an inward shift of the peak with increasing Reynolds number. A similar argument applies to the wall-normal velocity fluctuations which can be observed to be higher, close to the lower wall. Here, v_{rms}/U_w and w_{rms}/U_w are slightly greater than those of the long term averaged flow field. The converse is the case close to the upper wall. As the duct's centre is approached, u_{rms}/U_w drop to lower values than in the channel while v_{rms}/U_w and w_{rms}/U_w are very similar to those in the channel.

Next, the invariants of the non-dimensional Reynolds stress anisotropy tensor given by:

$$II = b_{ij}b_{ji}/2, \qquad III = b_{ij}b_{jk}b_{ki}/3, \qquad (6.4)$$

where

$$b_{ij} = \overline{u_i' u_j'}/\overline{u_k' u_k'} - \frac{1}{3}\delta_{ij}, \qquad (6.5)$$

will be examined. All possible turbulence states must lie within the Lumley triangle [13], bounded by the lines: $-II/3 = (III/2)^{2/3}$, $-II/3 = (-III/2)^{2/3}$ and $-II = 3(1/27 + III)$, corresponding to axisymmetric expansion, axisymmetric contraction and two-component turbulent states, respectively [14], while the bottom, left and right vertices represent the three-component isotropic, two-component axisymmetric and one-component limits, respectively. It should be noted that these

Fig. 6.12 Anisotropy invariant maps along lines parallel to the sidewalls: **a** at $z/h = 1$ (data set is shifted to the left) **b** Plane Couette flow data of Avsarkisov et al. [11]. Arrows indicate direction of increasing y

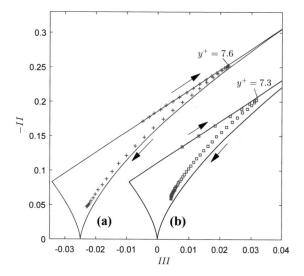

do not relate to the shape of the turbulent eddies; rather, they give an indication of the state of the Reynolds stress tensor. Figure 6.12 shows the anisotropy invariant map (AIM) for the long-time-averaged flow field as well as the plane Couette flow data of Avsarkisov et al. [11] at $Re_\tau = 125$. Data for only half of the duct is shown. Along the bisector of the moving walls (see Fig. 6.12a), the AIM is not very different from the one in plane Couette flow (Fig. 6.12b). Close to the wall, due to the damping of wall-normal velocity fluctuations, the turbulence is two-dimensional but at $y^+ \approx 7.6$ ($y^+ \approx 7.3$ in the plane Couette data), there is a switch to an axisymmetric state. As the centre of the duct is approached, the turbulence becomes increasingly isotropic due to the closeness of the fluctuations in all velocity components (see Fig. 6.10b). The level of isotropy at the centre is however higher in the square duct.

The variation of shear stress along the moving walls while the flow is in state A, as well as the wall-shear-stress distribution in long-time-averaged flow field are shown in Fig. 6.13a. The plots have been normalised by the long-term mean at the centre of the moving wall, $\bar{\tau}^*$. Close to the corners, the stresses increase dramatically[2] (see inset of Fig. 6.12a). Slow-moving fluid at the stationary side walls interact with fast-moving fluid at the moving wall, resulting in large velocity gradients. At the lower wall, the shear stress drops to a local minimum at $z/h = 0.45$ before approaching a local maximum at $z/h = 1$, while at the upper wall, a local minimum is found at the centre. As the flow switches to state B, the profiles become interchanged, hence no distinct extrema can be observed in the long-time-averaged flow field.

---

[2]It should be noted that there is a singularity at the corner. To check the effect on the computed turbulence statistics, results obtained in a duct of length 4 πh using two different meshes having 96 × 96 and 128 × 128 cells in duct's cross-section were compared and the difference in the turbulence statistics was found to be less than 1%. Hence the singularity can be said to have only a marginal effect on the turbulence statistics presented here, at least in regions which are away from the corners.

Fig. 6.13 Wall shear stress distribution at the moving walls. The flow is in state A: ——, lower wall; _ _ _ _, upper wall; -.-.-.-.-, long-time-averaged flow field. Inset shows the wall shear stress distribution close to a corner. Only half of the duct is shown due to symmetry about the moving-wall bisector

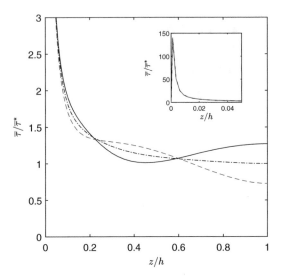

Figure 6.14a and b show the typical instantaneous velocity fields at x-z planes close to the lower and upper walls in the bi-stable states (In this case, the data for state A is presented). As earlier shown, Re_τ for this flow is as low as 70. This corresponds to a duct width of about 140 in wall units. Given that the separation distance of coherent structures in a turbulent flow is about 100 wall units [15], the duct is just barely wide enough to host the structures required to maintain a turbulent state. The figures reveal the existence of only a single high-speed streak centrally positioned along the upper wall and two at the lower wall, close to the corners. Hence the switching between states occurs in such a way as to increase the number of structures close to a given wall, to ensure continued sustenance of turbulence. The position of the streaks correspond to the wall shear stress extrema locations. To clearly show them, the data at $y/h = 0.2$ and $y/h = 1.8$, corresponding to locations of large u_{rms} are presented. The structures are persistent over the length of the duct, indicating that the states are very stable in the streamwise direction. Associated with each streak is a pair of counter-rotating vortices. Figure 6.14c shows a snapshot of the instantaneous velocity field in the cross-sectional plane of the duct at $x/h = 20$ where the two streaks in Fig. 6.14b are clearly separated. The high-speed streaks, indicated by mushroom-shaped velocity contours, can be observed to be positioned between the vortices. They are formed due to the lift-up of high speed flow by the vortices from the wall. These coherent structures have been shown to play a crucial role in the turbulence regeneration cycle [16]. When averaged in space and time, they result in the observed secondary flow pattern in a square duct.

Figure 6.15 shows the results of quadrant analysis carried out to determine the contribution of various turbulent events to the Reynolds shear stress, $\overline{u'v'}$. The Reynolds shear stress can be divided into four quadrants (Q1–Q4) depending on the sign of velocity fluctuations (see inset of Fig. 6.15a). In the analysis, the coordinate system

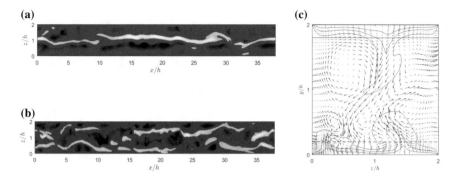

Fig. 6.14 Coherent structures. **a** and **b** are typical instantaneous velocity fields (not to scale) at $x - z$ planes located at $y/h = 1.8$ and 0.2, respectively, corresponding to regions of large fluctuations in streamwise velocity ; white contours indicate $u' > 0$ and black, $u' < 0$. Only fluctuations larger than 40% of the maximum in each plane are shown. **c** is the velocity field in the duct's cross-section at $x/h = 20$. $_ - _ - _ -$, $y/h = 0.2$; ———, $y/h = 1.8$. Flow is in state A

was modified such that positive v always pointed away from the nearest moving wall, while u was always positive in the translation direction of the nearest moving wall. In state A, along the bisector of the upper wall, Q1 events representing the ejection of fast-moving fluid away from the wall by positive wall-normal velocity fluctuations dominate. This is also the case at the lower wall, close to the corners (see Fig. 6.15b, in which the data along the line $z/h = 0.45$ passing through a region where the secondary flow vectors are pointed away from the wall is presented). However, along the bisector of the lower wall, Q3 events representing sweeping motion during which slow moving fluid is conveyed to the wall by negative wall-normal fluctuations account for the bulk of $\overline{u'v'}$ generation (see Fig. 6.15c). A closer examination of the velocity fields in Fig. 6.4a reveals that along the bisector of the upper wall (where quadrant analysis shows ejection events to dominate), the secondary flow can be observed to transport momentum away from the wall, causing the axial velocity contour to bulge towards the duct's interior. Similarly, at the lower wall (where sweeeping motion dominates), the secondary flow transports momentum towards the wall, causing the contours of axial velocity to bulge inwards. It is thus evident that the secondary motion is closely related to the near-wall ejection and sweeping events.

6.4   Summary

Direct numerical simulations of turbulent Couette flows in a square duct at relatively low Reynolds numbers have been carried out. The flows are driven by a pair of opposite counter-moving walls translating with the same speed, resulting in a zero net transport of fluid through the duct. A turbulent state was found to be maintained

Fig. 6.15 Quadrant analysis of the Reynolds shear stress, while the flow is in state A: **a** along the bisector of the upper wall, **b** at $z/h = 0.45$ (line passes through a region where the secondary flow vectors are pointed away from the lower wall). $+$, Q1; -.-.-, Q2; $\times$, Q3; ——— , Q4; - - - - -, $\overline{u'v'}$ total **c** along the bisector of the lower wall. $+$, Q1; -.-.-, Q2; $\times$, Q3; ———, Q4; - - - - -, $\overline{u'v'}$ total

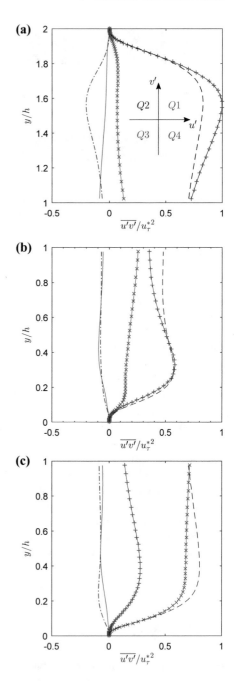

only for Reynolds numbers greater than about 875. This is much higher than the $Re_c \approx 325 - 375$ observed in plane Couette flow studies, [6–8, 17] thus highlighting the stabilising effect of the side walls on the flow. At Reynolds numbers close to the critical, the flow was observed to exist in two states, one being a mirror reflection of the other, with symmetry about the common bisector of the moving walls. In these states, the secondary motion (which is different from that in Poiseuille flow) is characterised by four vortices, induced by gradients of the anisotropic Reynolds normal stresses, $\overline{w'^2} - \overline{v'^2}$ and cross stream shear stress, $\overline{v'w'}$. Due to the intermittency in the flow, large fluctuations in the velocity, about the long-term mean were observed. Instantaneous flow field visualisations reveal the existence of coherent structures which are persistent over the length of the duct, thus indicating that the states are very stable in the streamwise direction. Quadrant analysis at different locations in the duct indicate that the secondary flow is closely related to the near-wall ejection and sweep events, ejection dominating in regions where the secondary flow transports momentum away from the wall and sweeps dominating in regions where momentum is transported towards the wall.

References

1. Hsu HW, Hsu JB, Lo W, Lin CA (2012) Large eddy simulations of turbulent Couette-Poiseuille and Couette flows inside a square duct. J Fluid Mech 702:89101
2. Lo W, Lin CA (2006) Mean and turbulence structures of Couette-Poiseuille flows at different mean shear rates in a square duct. Phys Fluids 18(6):068103
3. Uhlmann M, Pinelli A, Kawahara G, Sekimoto A (2007) Marginally turbulent flow in a square duct. J Fluid Mech 588:153–162
4. Jiménez J, Moin P (1991) The minimal flow unit in near-wall turbulence. J Fluid Mech 225:213–240
5. Hamilton JM, Kim J, Waleffe F (1995) Regeneration mechanisms of near-wall turbulence structures. J Fluid Mech 287:317–348
6. Daviaud F, Hegseth J, Bergé P (1992) Subcritical transition to turbulence in plane Couette flow. Phys Rev Lett 69:2511–2514
7. Tillmark N, Alfredsson PH (1992) Experiments on transition in plane Couette flow. J Fluid Mech 235:89–102
8. Shi L, Avila M, Hof B (2013) Scale invariance at the onset of turbulence in Couette flow. Phys Rev Lett 110(20):204502
9. Vinuesa R, Schlatter P, Nagib HM (2015) On minimum aspect ratio for duct flow facilities and the role of side walls in generating secondary flows. J Turbul 16(6):588–606
10. Abe H, Kawamura H, Matsuo Y (2001) Direct numerical simulation of a fully developed turbulent channel flow with respect to the Reynolds number dependence. J Fluids Eng 123(2):382–393
11. Avsarkisov V, Hoyas S, Oberlack M, Garca Galache JP (2014) Turbulent plane Couette flow at moderately high Reynolds number. J Fluid Mech 751:R1
12. Pirozzoli S, Bernardini M, Orlandi P (2014) Turbulence statistics in Couette flow at high Reynolds number. J Fluid Mech 758:327–343
13. Lumley JL (1979) Computational modeling of turbulent flows. Adv Appl Mech 18:123–176
14. Simonsen AJ, Krogstad PÅ (2005) Turbulent stress invariant analysis: clarification of existing terminology. Phys Fluids 17(8):088103

15. Kim J, Moin P, Moser R (1987) Turbulence statistics in fully developed channel flow at low Reynolds number. J Fluid Mech 177:133–166
16. Jiménez J, Pinelli A (1999) The autonomous cycle of near-wall turbulence. J Fluid Mech 389:335–359
17. Lundbladh A, Johansson AV (1991) Direct simulation of turbulent spots in plane Couette flow. J Fluid Mech 229:499516

Chapter 7
Turbulent Duct Flows with Polymer Additives

7.1 Introduction

Long-chain flexible polymers of high molecular weight are known to greatly reduce the skin friction drag in turbulent flow when added to a liquid even at minute concentrations [1]. This phenomenon has prompted a large number of studies see e.g. [2–4], yet it has never been possible to relate the degree of drag reduction (DR) to a measurable fluid property. A good understanding of their extensional rheological properties is believed to be crucial to obtaining a better insight into their DR mechanism. Until fairly recently, with the introduction of the Capillary Break-up Extensional Rheometer (CaBER) [5, 6], these properties have been elusive to experimentalists (due to the dilute nature of the polymer solutions). Mechanical degradation of polymer molecules is also a major issue, giving rise to poor repeatability and large uncertainties in experimental data. In this chapter, the turbulent DR mechanism in flow through ducts of circular, rectangular and square cross-sections (each commonly-encountered geometries in the study of wall-bounded turbulence) is investigated using two grades of polyacrylamide in aqueous solution having different molecular weights and various semi-dilute concentrations. Specifically, the relationship between drag reduction and fluid elasticity is explored, the mechanical degradation of polymer molecules purposely exploited to vary their rheological properties. Time-resolved velocity measurements for various DR levels are also obtained with particle image velocimetry and laser Doppler velocimetry. Elasticity is quantified via relaxation times determined from uniaxial extensional flow using CaBER. For the first time, it is shown that quantitative predictions of DR in a range of shear flows can be made from a single measurable material property of a polymer solution, independent of other experimental variables, at least for this flexible linear polyacrylamide.

© Springer Nature Switzerland AG 2019
B. Owolabi, *Characterisation of Turbulent Duct Flows*,
Springer Theses, https://doi.org/10.1007/978-3-030-19745-2_7

7.2　Working Fluid Preparation and Rheological Characterisation

As explained in Chap. 3, two grades of polyacrylamide solutions were studied: FloPAM AN934SH ("PAA") and Separan AP273E ("Separan") both supplied by Floreger. These polymer solutions do degrade under high shear, hence great care was taken in their preparation. First, stock solutions with concentrations ranging from 1200–1500 ppm were manually prepared outside the rigs and left for a period of twenty four hours to homogenise before being diluted to the desired concentrations in situ by mixing with water at very low pump speeds. For every run in the drag-reduction studies, the rheological properties of the polymer solutions were closely monitored. Fluid samples were taken at different pumping times for viscosity and relaxation time measurements using an Anton Paar MCR302 controlled-stress torsional rheometer and a CaBER respectively, in conjunction with pressure-drop measurements to estimate the level of drag reduction. For a detailed explanation of the operating principle of the CaBER, the reader is referred to Sect. 3.5.2.

Example data sets of shear viscosity and the variation of relaxation time with pumping time are shown in Fig. 7.1. Polymer degradation can be observed to bring about a large reduction in the zero-shear viscosity, the sharpest drop occurring in the first 45 min of pumping (as shown in Fig. 7.1a). Power-law fits to the shear viscosity curves (inset of Fig. 7.1a) also show a decrease in the amount of shear thinning during this time frame. At longer times, the rate of polymer degradation slows down and shear viscosity becomes roughly constant. The relaxation times measured by CaBER (λ_c) can be observed to follow a similar trend (see Fig. 7.1b), dropping from relatively large values at the start of pumping before asymptoting to a value of about 4 ms at long times (close to the minimum resolution of the standard CaBER technique). From the log-log plot (inset of Fig. 7.1b), it can be observed that λ_c scales roughly as $t^{-0.5}$, where t is the pumping time in seconds.

7.3　Mean Streamwise Velocity Measurements

Figure 7.2 shows the velocity profiles in the pipe, rectangular channel and square duct at different levels of drag reduction. In this study, DR is defined as:

$$\% DR = \left. \frac{\overline{\Delta p_N} - \overline{\Delta p_V}}{\overline{\Delta p_N}} \times 100 \right|_{\dot{m}} , \tag{7.1}$$

where $\overline{\Delta P_N}$ and $\overline{\Delta P_V}$ are the mean pressure drops in the Newtonian fluid and polymer (viscoelastic) solution respectively, measured at the same mass flow rate, $\dot{m}$. The time-resolved velocity data were obtained using LDV in the rectangular and square channels and by stereoscopic PIV (SPIV) in the pipe. With SPIV, it was possible to

Fig. 7.1 Variation of the rheological properties of polymer solutions with pumping time t. **a** Example shear viscosity curves for PAA (concentration of 250 ppm in the square duct) at various stages of mechanical degradation: a power-law model ($\mu = K\dot{\gamma}^{n-1}$) has been fitted in the range shown by the two vertical dashed lines and the inset shows the change in power law index, n, with pumping time t. **b** Typical relationships between CaBER-measured relaxation time and pumping time showing rapid degradation, with an inset of the same data on log-log axes highlighting $\lambda_c \sim t^{-0.5}$

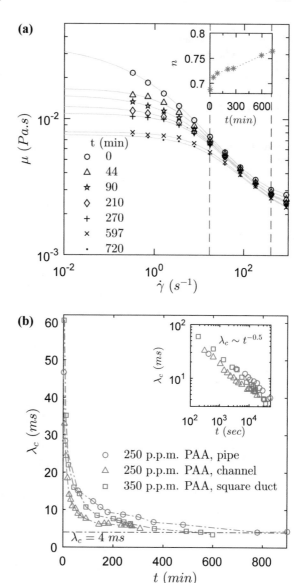

monitor the changes in the mean velocity profile with decreasing drag-reduction level in a single experiment, since the technique provides velocity information in the entire cross section of the pipe within a reasonable time-frame over which degradation is minimal. With LDV (a point-wise measurement technique) however, useful velocity data could only be obtained after long pumping times when the polymer rheological properties were not changing much and $\%DR$ was roughly constant.

Fig. 7.2 Time-averaged
streamwise velocity profiles
for 250 ppm PAA solutions
at various levels of drag
reduction in three parallel
shear flows. **a** Azimuthally
averaged data obtained using
SPIV in the cylindrical pipe,
b LDV data from the
rectangular channel **c** LDV
profiles along the wall
bisector of the square duct

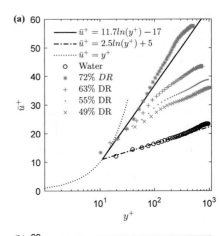

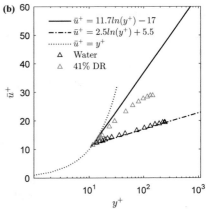

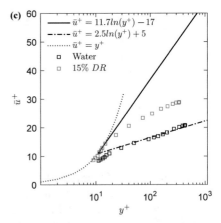

From the plots shown in Fig. 7.2, a thickened buffer layer can be clearly seen, the buffer layer extending across the entire cross section at MDR (72% DR in the pipe, see Fig. 7.2a). The Weissenberg number,[1] Wi, for all the polymer data was greater than 1, indicating the importance of polymer stretching ($Wi = \lambda_c \dot{\gamma}$, where $\dot{\gamma}$ is the mean shear rate at the wall). In the pipe, the profile of $\bar{u}^+ = \bar{u}/u_\tau$ lies slightly above Virk's MDR line, $\bar{u}^+ = 11.7ln(y^+) - 17$, for $y^+ > 80$ (y^+ is the distance from the wall normalised by the viscous length, ν/u_τ, where ν is the kinematic viscosity at the wall shear rate). It is however within the 95% confidence interval of the Virk profile [7]. In fact, it is well known [8, 9] that data nominally following Virk's MDR asymptote is not truly logarithmic. This can be easily verified by computing an indicator function

$$\Phi = y^+ \frac{d\bar{u}^+}{dy^+}, \tag{7.2}$$

whose constancy when plotted against y^+, shows the presence of a truly logarithmic velocity distribution. (in a logarithmic region, Φ takes the value of the log-law slope, $1/\kappa$).

Plots of Φ versus y^+ for the pipe flow data of Fig. 7.2a are presented in Fig. 7.3a. While a plateau can be observed for water in the region $80 \lesssim y^+ \gtrsim 230$, where $\Phi = 1/\kappa$ is equal to the well-known value of 2.5 for Newtonian fluids (indicated as the dash-dotted line), the data at 72% DR (MDR) shows no region of constant Φ. Hence the MDR equation should be viewed as an idealisation which is helpful in highlighting that there exists a parameter regime where the velocity profile is only weakly sensitive to polymer and flow properties.

As $\%DR$ reduces due to polymer degradation, a decrease in the values of $\bar{u}^+$ can be observed (see Fig. 7.2a), the velocity profiles smoothly interpolating between the MDR line and von Karman log law for Newtonian fluids up to some y^+ before becoming roughly parallel to the Newtonian log law (although this simple view is known to fail at high drag-reduction where the low-law slope is no longer observed [10]). A similar trend can be observed in the rectangular and square channels (see Figs. 7.2b, c, 7.3b and c; the scatter in the indicator function plots can be attributed to the use of first order forward differences in estimating $d\bar{u}^+/dy^+$, as the data points are not uniformly spaced).

7.4 Instantaneous Velocity Fluctuations

Figure 7.4 shows the instantaneous streamwise velocity fluctuations, u' in the cylindrical pipe at different %DR levels. These fluctuations have been computed by subtracting the azimuthally averaged (also time-averaged) velocities at different radial

---

[1]$Wi = \lambda_c \dot{\gamma}$. Neglecting shear thinning effect, the shear stress at the wall can be expressed as, $\tau_w = \mu \dot{\gamma}$. The Weissenberg number can thus be written as $Wi = \frac{\lambda_c u_\tau^2}{\nu} = \lambda_c^+ u_\tau$, where $u_\tau = \sqrt{\tau_w/\rho}$ is the friction velocity.

Fig. 7.3 Indicator function, $\Phi = y^+ d\bar{u}^+/dy^+$, at various drag reduction levels: **a** cylindrical pipe **b** rectangular channel. ——, $1/\kappa = 11.7$; -.-.-, $1/\kappa = 2.5$. Symbols and colours as in Fig. 7.5a, b, c square duct. ——, $1/\kappa = 11.7$; -.-.-, $1/\kappa = 2.5$. Symbols and colours as in Fig. 7.2c

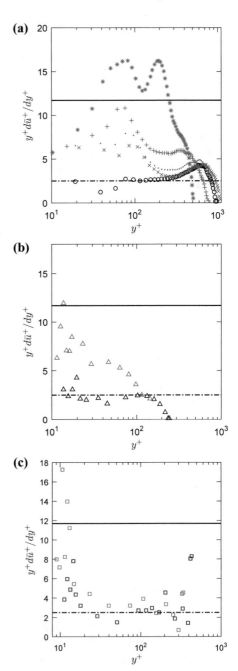

locations, r_o, from the instantaneous velocity fields at those locations (the locations at which velocity data are presented correspond to $y^+ \approx 170$). The quasi three-dimensional flow fields were reconstructed from 2-D snapshots of velocity obtained from time-resolved SPIV measurements by applying Taylor's hypothesis of frozen turbulence [11]. The hypothesis states that "if the velocity of the airstream which carries the eddies is very much greater than the turbulent velocity, one may assume that the sequence of changes in u at the fixed point are simply due to the passage of an unchanging pattern of turbulent motion over the point". Mathematically, this can be formulated as see [12]

$$u(x, t) = u(x - \mathcal{U}_c \Delta t, t + \Delta t), \tag{7.3}$$

where $\mathcal{U}_c$ is the velocity at which the flow structures are being convected (which in this study, is taken as the bulk velocity, U_b), Δt is the time delay between measurements and x is the streamwise co-ordinate. Since the velocity fluctuations are about one order of magnitude less than U_b and the sampling rates in the SPIV measurements are high (greater than 600 Hz), Taylor's hypothesis can be accurately used to convert the temporal data to spatial data. The approach has been successfully used to reproduce the large scale motions in wall bounded turbulent flows see e.g. [13].

An increase in the length and width of the flow structures can be observed with increasing level of drag reduction, the structures becoming more coherent. The small turbulence scales can also be observed to be attenuated, hence at MDR (72% DR) there is very little scale separation. The mechanism for the production and dissipation of turbulence is thus expected to be very different than in the Newtonian case. These results are consistent with findings in the literature see [4, 10]; however, to the best of the author's knowledge, the present study is the first in which the evolution of the flow structures from MDR to zero drag reduction has been monitored in an experiment.

7.5 Universal Relationship Between Drag Reduction and Fluid Elasticity

The Fanning friction factors (f) computed from pressure-drop measurements are shown in Fig. 7.5. The measurements have been taken at different pumping times and mass flow rates and cover a wide range of %DR. f is plotted against the Reynolds number, $Re = \rho U_b h / \mu$, where ρ, U_b, h and μ represent the density, bulk velocity, duct half height (or radius in the case of the pipe) and apparent viscosity at the wall shear rate at any given t, respectively. The Reynolds number ranges studied varied by roughly an order of magnitude in each geometry; $10^3 - 10^4$ for the channel, $4 \times 10^3 - 2 \times 10^4$ for the square duct and $5 \times 10^3 - 5 \times 10^4$ for the pipe. The approximate mean wall shear rate, estimated from shear viscosity curves, is the shear rate corresponding to the shear stress, τ_w computed from pressure-drop measurements at time, t. This procedure is possible as the rheometer provides the link

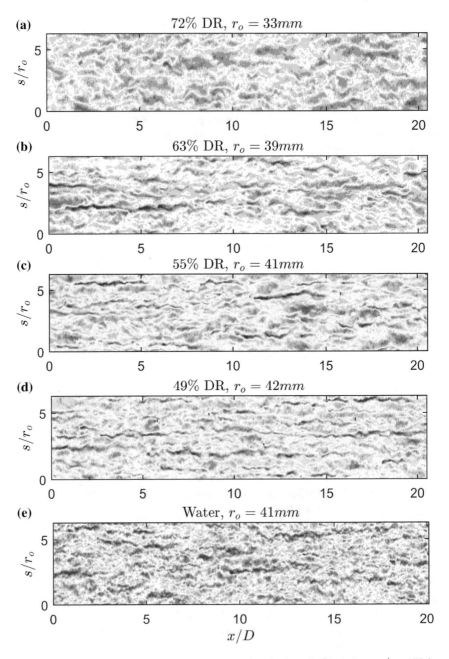

Fig. 7.4 Contour plots of streamwise velocity fluctuations in the cylindrical pipe at $y^+ \approx 170$ for varying levels of %DR. Blue, $u' < 0$; red, $u' > 0$; white, $u' = 0$. D is the pipe diameter (=100 mm) and s, the azimuthal distance at a given radial location, r_0. Plots correspond to the data in Fig. 7.2a. It should be noted that as the polymer degrades, the wall shear stress changes hence the radial location corresponding to a fixed y^+ also changes

Fig. 7.5 Fanning friction factors at different flow rates and pumping times for **a** cylindrical pipe **b** rectangular channel. Dotted lines ($\cdots$) are the appropriate laminar flow equations for Fanning friction factor; dot-dashed lines ($-\cdot-$) are the correlations of Blasius (for pipe and square duct) and [14] (for rectangular channel); Solid lines ($—$) are the correlations of [2] (for pipe and rectangular channel) and [15] (for square duct) at MDR, the shaded regions represent $f = \pm 10\%$ of MDR. $Re_\delta = \rho U_b D_h / \mu$ where D_h is the hydraulic diameter **c** square duct. Legend as in (**a**) and (**b**)

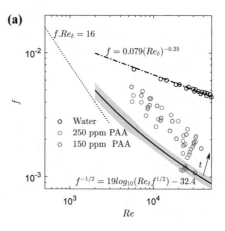

(a)

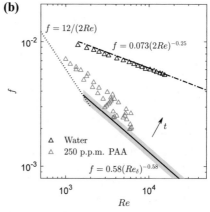

(b)

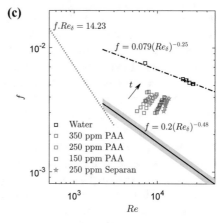

(c)

between shear stress and shear rate through the measured shear viscosity. It should be noted that the mean shear rate determined in this manner is merely an approximation for shear-thinning fluids and is exact only in the Newtonian limit [16]. As expected, the polymer data lie between Virk's MDR and the correlation of Blasius or [14] for Newtonian turbulent flow, f approaching the Newtonian values with increasing pumping time.

In their present form, Figs. 7.1 and 7.5 have very limited practical applicability in terms of predictive capability. It is therefore desirable to collapse the data for all three geometries onto a single curve. As a step towards this goal, $\%DR$ is plotted against the Weissenberg number, $Wi = \lambda_c \dot{\gamma}$ in Fig. 7.6b. Quite remarkably, given the large range of Reynolds number, polymer concentration and different geometries a good data collapse can be observed. The data is reasonably well fit (R-squared $=$ 0.82) by the following equation:

$$\%DR = 2C_1 \left(\frac{1}{1 + e^{Wi_c - Wi}} - Wi_c \right), \tag{7.4}$$

where $C_1 = 64$ is the approximate limiting value of $\%DR$ as $Wi \to \infty$ and Wi_c is the critical Weissenberg number for onset of drag-reduction (set to be $Wi_c = 0.5$). Hence, the degree of drag-reduction has been related to a single measurable extensional rheological property of the polymer solutions independent of flow geometry, concentration, degradation and other experimental variables. The onset of $\%DR$ can be observed to take place at $Wi \approx 0.5$ which, although in agreement with the theory of [17] i.e. potentially related to the so-called "coil-stretch" transition, may just be fortuitous as Wi here is based on the average wall shear rate rather than the fluctuating strain rates or the largest Lyapunov exponent [18]. MDR is attained at $Wi \gtrsim 5$ for which $\%DR$ becomes independent of Wi.

A strong qualitative agreement between the universal form of the experimental data presented here with DNS results obtained from various models shown in [19] is noted: however the DNS data requires an additional fitting parameter ("LDR") which represents the asymptotic limit of large Weissenberg numbers which is not known a priori. In contrast, Eq. 7.4 enables the degree of drag reduction at a given flow rate in a given geometry to be estimated, using an iterative procedure, solely if the polymer shear rheology and relaxation time are known. An example of such iterative procedure is as follows. Firstly an initial estimate of $\overline{\Delta p_V}$ can be obtained from the correlation of Blasius or [14] (using viscosity of solvent), then this pressure drop can be used to determine the mean wall shear stress from a momentum balance (i.e. $\bar{\tau}_w = \overline{\Delta p} D / 4\ell$ in a pipe, for example); from this shear stress the wall shear rate can then be obtained using the rheology data (from a plot of shear stress versus shear rate). This wall shear rate multiplied by the CaBER relaxation time gives the Weissenberg number; now, $\%DR$ can be obtained from Eq. 7.4 and a new value of $\overline{\Delta p_V}$ determined from Eq. 7.1. This procedure is repeated until $\overline{\Delta p_V}$ converges. Alternatively, one could use the average shear rate (U_b / h) to estimate Wi and then obtain $\%DR$ from Eq. 7.4 which will enable $\overline{\Delta p_V}$ to be determined. This pressure

Fig. 7.6 a Combined f-Re data for cylindrical pipe, rectangular channel and square duct (symbols and colours as in Fig. 7.5). **b** Variation of %DR with Weissenberg number. Solid black line represents the new correlation:
$$\%DR = 2C_1(1/[1 + e^{Wi_c - Wi}] - Wi_c)$$

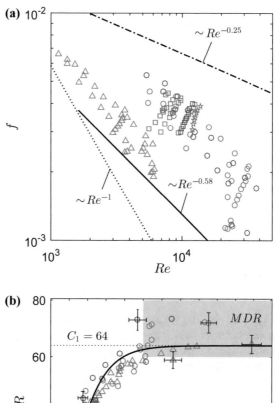

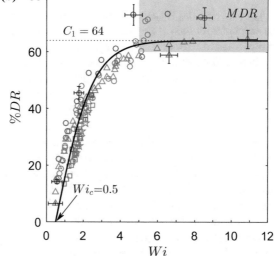

drop could then be used to determine the mean wall shear stress etc. in a similar manner.

In observing a working functional dependence of %DR on Wi alone, the dependence on solvent-to-total viscosity ratio β, on inertia (i.e. Reynolds number) and on other viscometric functions, e.g. first or second normal-stress differences, is neglected. As all of the concentrations are quite similar, all being within the semi-dilute regime, the effect of β must contribute to the spread of the data around the fit.

Fig. 7.7 Reynolds number
dependence of $\%DR$ at
MDR

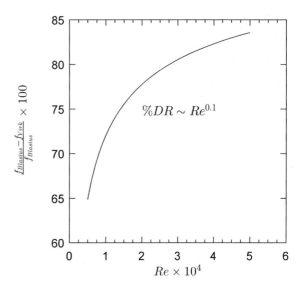

It is noted that the variation in the data is larger than the experimental uncertainty associated with the data which is estimated to be $\pm 3 - 3.5\%$ as highlighted by the representative error bars shown in Fig. 7.6b.

In the MDR limit it is also known that there remains some weak Re dependence, as $\%DR$ scales roughly as $Re^{0.1}$ in this limit (see Fig. 7.7), and, as already discussed, there is significant spread in the literature for data nominally at MDR [7] as is highlighted by the grey region in Fig. 7.6b. Given the quality of the data collapse illustrated in Fig. 7.6b, both β and Re seem to be second-order effects, at least for the concentrations (on the order of c^*) and range of Re ($\approx 10^3 - 5 \times 10^4$) studied for this particular flexible polymer. Although the data, and the correlation proposed in Eq. 7.4, suggest that no critical polymer concentration for drag reduction onset is required (provided the mean wall shear rate is sufficient to make $Wi \gtrsim 0.5$), it should be noted that the range of β probed here (≈ 0.1) means the polymeric viscosity (and hence polymeric stress) is never negligible. In the limit where the solvent contribution to the stress dominates (i.e. $\beta \to 1$), simulations suggest a critical concentration is required for onset regardless of shear rate [20].

No dependence on the Trouton ratio Tr (i.e the ratio of extensional viscosity to shear viscosity) is observed but in this study, Tr and λ_c cannot be independently adjusted as both are related to the length of the polymer molecule. Finally the data collected in [7] for different geometries—boundary-layer, pipe and channel-flow— also exhibits a similar spread in data nominally at MDR suggesting that some of the spread here could also be related to the use of different geometries and the simplistic estimate of the mean wall shear rate. In particular the square duct, and to a lesser extent the channel, exhibits a mean wall shear rate that varies around its periphery whereas the pipe is axisymmetric and therefore spatially constant.

7.6 Summary

Turbulent drag reduction with polyacrylamide, a flexible polymer, has been investigated in a cylindrical pipe, rectangular channel and square duct. The polymer solutions were subjected to various levels of degradation by recirculating through the experimental rigs and their shear and extensional rheological properties closely monitored. Polymer relaxation times were observed to scale roughly as $t^{-0.5}$. Profiles of streamwise velocity ($\bar{u}^+$) at various levels of drag-reduction indicated a thickening of the buffer layer as observed in previous experimental studies, the buffer layer extending across the entire cross section at MDR. A plot of $\%DR$ against Weissenberg number was found to collapse the data, with the onset of DR occurring at $Wi \approx 0.5$ and MDR at $Wi \gtrsim 5$, thus allowing for an a priori quantitative prediction of drag reduction from a knowledge of polymer relaxation time, flowrate and geometric lengthscale (using either the Newtonian pressure drop combined with rheology data, or the average shear rate as an initial guess to determine Wi, combined with an iterative procedure).

References

1. Toms BA (1948) Some observations on the flow of linear polymer solutions through straight tubes at large Reynolds numbers. In: Proceedings of the 1st international congress on rheology, vol. 2, pp. 135–141
2. Virk PS (1975) Drag reduction fundamentals. AIChE J 21(4):625–656
3. Den Toonder JMJ, Draad AA, Kuiken GDC, Nieuwstadt FTM (1995) Degradation effects of dilute polymer solutions on turbulent drag reduction in pipe flows. Appl Sci Res 55(1):63–82
4. White CM, Mungal MG (2008) Mechanics and prediction of turbulent drag reduction with polymer additives. Ann Rev Fluid Mech 40:235–256
5. Bazilevsky A, Entov V, Rozhkov A (1990) Liquid filament microrheometer and some of its applications. In: Third European rheology conference and golden jubilee meeting of the British society of rheology, pp. 41–43. Springer, Dordrecht
6. Rodd L, Scott TP, Cooper-White JJ, McKinley GH (2005) Capillary break-up rheometry of low-viscosity elastic fluids. Appl Rheol 15:12–27
7. Graham MD (2014) Drag reduction and the dynamics of turbulence in simple and complex fluids. Phys Fluids 26(10):101301
8. Elbing BR, Perlin M, Dowling DR, Ceccio SL (2013) Modification of the mean near-wall velocity profile of a high-reynolds number turbulent boundary layer with the injection of drag-reducing polymer solutions. Phys Fluids 25(8):085103
9. White C, Dubief Y, Klewicki J (2012) Re-examining the logarithmic dependence of the mean velocity distribution in polymer drag reduced wall-bounded flow. Phys Fluids 24(2):021701
10. Warholic MD, Massah H, Hanratty TJ (1999) Influence of drag-reducing polymers on turbulence: effects of Reynolds number, concentration and mixing. Exp Fluids 27(5):461–472
11. Taylor GI (1938) The spectrum of turbulence. Proc R Soc Lond A 476–490
12. Townsend AA (1976) The structure of turbulent shear flow, 2nd edn. Cambridge University Press, New York
13. Dennis DJC, Nickels TB (2008) On the limitations of Taylor's hypothesis in constructing long structures in a turbulent boundary layer. J Fluid Mech 614:197–206
14. Dean RB (1978) Reynolds number dependence of skin friction and other bulk flow variables in two-dimensional rectangular duct flow. J Fluids Eng 100(2):215–223

15. Hartnett JP, Kwack EY, Rao BK (1986) Hydrodynamic behavior of non-Newtonian fluids in a square duct. J Rheol 30(4):S45–S59
16. Housiadas KD, Beris AN (2004) Characteristic scales and drag reduction evaluation in turbulent channel flow of nonconstant viscosity viscoelastic fluids. Phys Fluids 16(5):1581–1586
17. Lumley JL (1973) Drag reduction in turbulent flow by polymer additives. J Polym Sci Macromol Rev 7:263–290
18. Stone PA, Graham MD (2003) Polymer dynamics in a model of the turbulent buffer layer. Phys Fluids 15(5):1247–1256
19. Housiadas KD, Beris AN (2013) On the skin friction coefficient in viscoelastic wall-bounded flows. Int J Heat Fluid Flow 42:49–67
20. Lee D-H, Akhavan R (2009) Scaling of polymer drag reduction with polymer and flow parameters in turbulent channel flow. In: Advances in turbulence XII, pp. 359–362

Chapter 8
Conclusions and Recommendations

8.1 Introduction

In this chapter, a summary of the contributions made to the knowledge on the characteristics of turbulent flows in ducts is given. Suggestions for future research are also presented. The summary is organised into three main sections. In Sect. 8.2.1, findings on transition to turbulence in duct flows are given, while in Sect. 8.2.2 the results on "marginally" and fully-turbulent flows in a square duct are summarised. Finally, in Sect. 8.2.3, the main findings from the study on polymer flow in ducts are highlighted.

8.2 Summary of Findings

8.2.1 Transition to Turbulence in Duct Flows

Laminar/turbulent transition in both pressure-driven (Poiseuille) and wall-driven (Couette) flows in a square duct have been investigated. Experimental data in the former indicate that Coriolis effects due to the rotation of the earth can significantly affect the onset of turbulence, by distorting the laminar base flow, resulting in lower values of the critical Reynolds number. At $Ek \approx 7$, a limiting Reynolds number (Re) of about 940 for transition was observed. This is within the range obtained in previous numerical studies ([1]: $Re = 1077$; [2]: $Re = 865$). Direct numerical simulations of zero-net-flux wall-driven flow in a square duct indicate that a turbulent state can only be maintained for Reynolds numbers (Re_w) greater than about 875. This is much higher than the $Re_w \approx 325 - 375$ observed in plane Couette flow studies (see e.g. [3–6]), hence the side walls have a stabilising effect of the flow.

© Springer Nature Switzerland AG 2019
B. Owolabi, *Characterisation of Turbulent Duct Flows*,
Springer Theses, https://doi.org/10.1007/978-3-030-19745-2_8

8.2.2   Turbulent Duct Flows at Low Reynolds Numbers

Velocity measurements in a square duct, in purely pressure-driven flows ranging from "marginally turbulent" to fully turbulent, have been obtained. The results are in good agreement with the DNS data of [1] at $Re_\tau \approx 80$ and [7] at $Re_\tau \approx 160$. For the first time, the alternation of the flow field between two states, in the marginal turbulence regime, originally predicted by the DNS of [1], is verified experimentally. Probability density functions of streamwise velocity at certain distance from the duct wall were found to be bimodal, and the joint probability density functions of streamwise and wall-normal velocity also featured two peaks corresponding to each of the two states: one essentially unidirectional ($v^+ \approx 0$) at the measurement location and the other containing a significant secondary flow component ($v/\bar{u} \approx 0.03$). The integral time scale in marginally-turbulent flow was observed to be higher than that in fully-turbulent flow, indicating that the former is correlated over a larger time period, thus reinforcing the evidence that the flow remains in a particular state for a significant amount of time. By applying Taylor's hypothesis to the data, it was shown that there was also a spatial switching along the length of the duct.

Direct numerical simulations of purely wall-driven turbulent flow in a square duct at $Re_w \approx 1500$ also reveal an alternation in time between two states, thus indicating that the phenomenon is not unique to Poiseuille flows. For the simulations, the case in which a pair of opposite counter-moving walls translating with the same speed drives the flow was considered. The two states are each characterised by a four-vortex secondary flow pattern, one being a mirror reflection of the other, and the flow remains approximately symmetrical about the common bisector of the moving walls. Due to the intermittency, large velocity fluctuations about the long-term mean were observed at different locations in the duct, consistent with findings in turbulent Poiseuille flow at low Reynolds numbers. Instantaneous flow field visualisations reveal the existence of coherent structures which are persistent over the length of the duct, thus indicating that the states are very stable in the streamwise direction. Quadrant analysis of the Reynolds shear stress shows that the secondary motions are closely related to the near-wall ejection and sweep events, ejection dominating in regions where the secondary flow transports momentum away from the wall and sweeps dominating in regions where momentum is transported towards the wall.

8.2.3   Turbulent Duct Flows with Polymer Additives

Turbulent flow of dilute and semi-dilute solutions of polyacrylamide, a flexible polymer, in ducts of circular, rectangular and square cross-sections have been investigated. These fluids are highly drag reducing, but they degrade when subjected to a high level of deformation. The mechanical degradation of the polymer molecules was exploited to vary their rheological properties, with a view to obtaining a relationship between drag reduction and fluid elasticity. The latter was quantified via relaxation times determined from uniaxial extensional flow using a capillary breakup apparatus. Time-resolved velocity data at various levels of drag reduction were also obtained using LDV and SPIV.

The profiles of streamwise velocity indicate a thickening of the buffer layer, as observed in previous experimental studies, the buffer layer extending across the entire cross-section at maximum drag reduction. Quasi 3D flow fields reconstructed from 2D snapshots of velocity obtained from time-resolved SPIV measurements in the cylindrical pipe show an increase in the length scales of the turbulence structures and an attenuation of small scales at increasing levels of drag reduction. Skin-friction data from all the geometries are highly scattered and hence, of limited practical applicability in terms of predicting drag reduction. A plot of $\%DR$ against Weissenberg number was found to collapse the data, with the onset of DR occurring at $Wi \approx 0.5$ and MDR at $Wi \gtrsim 5$. The data are reasonably well fit ($R^2 = 0.82$) by the following equation: $\%DR = 2C_1(1/[1 + e^{Wi_c - Wi}] - Wi_c)$, where $C_1 = 64$ is the approximate limiting value of $\%$DR as $Wi \to \infty$ and Wi_c is the critical Weissenberg number for the onset of drag reduction. Hence for the first time, quantitative predictions of drag reduction in a range of shear flows can be made from a single measurable material property of a polymer solution—the uniaxial extensional relaxation time.

8.3 Suggestions for Future Work

The turbulent flow of Newtonian and non-Newtonian fluids in ducts have been extensively investigated and a clearer picture of their characteristics have been obtained. However, some questions still remain unanswered, hence there is room for further research. In the following sections, some recommendations for future work are given.

8.3.1 Transition to Turbulence and Marginally Turbulent Flows

Exact coherent structures, such as travelling waves, are known to play important roles in the transition dynamics in ducts. These alternative solutions to the Navier-Stokes equations have been discovered in computational studies on transition in channels, pipes and square ducts (see e.g. [8–11]). Their existence have also been experimentally confirmed in pipe flows [12]. However, there are no experimental data to verify their existence in other geometries. It is suggested that an approach similar to that of [12] be employed in the square duct and rectangular channel to detect these structures. With regards to "marginally turbulent" duct flows where there is a switching between two states, it is suggested that proper orthogonal decomposition (POD) be carried out to extract the most energetic structures.

8.3.2 Turbulent Couette-Poiseuille Flows

A new test-section for the investigation of Couette-Poiseuille flows have been constructed and tested in the laminar regime (see Appendix A). The next step would be to perform turbulent flow experiments. Given the finite size of the experimental facility, an investigation of flow development in the turbulent regime will be helpful in determining the range of Reynolds numbers at which fully-developed turbulent flows can be obtained. In evaluating the development length, velocity profiles at different streamwise distances can be compared with the large eddy simulation data of [13]. It will be interesting to investigate the effectiveness of wall motion as a flow control mechanism. In this regard, the transition to turbulence at different ratios of wall to bulk velocity can be studied and the observed critical Reynolds numbers compared with that in purely pressure-driven flow. Similarly, modifications to the secondary flow can be investigated. For this, SPIV measurements will be very useful. The SPIV technique provides information about the velocity field over an entire cross-section, thus making it easy for the cross-stream structures to be captured. Besides translating the wall in the same direction as the bulk flow, it is recommended that wall motion also be imposed in the opposite direction, to allow for the study of reversed flows, flow separation, and the resulting effects on the wall shear stress. The drive train for the belt can also be programmed to generate other motions (such as oscillatory) and the resulting flow fields, investigated. Finally, it is proposed that an experimental verification of large scale structures in turbulent Couette flow be conducted. DNS of these flows indicate that the structures are highly persistent in space and time.

8.3.3 Polymer Flows

Available Newtonian DNS data at Reynolds numbers ranging from marginally turbulent to fully turbulent show that the secondary flow in a square duct is related to the wall shear stress (τ_w) distribution. This is also indicated by the recent DNS of [14] in polymer flows using FENE-P model; however, there are no experimental data to validate these findings. In the present study, an attempt to determine the distribution of τ_w at high values of drag reduction was unsuccessful, because the degradation of polymer molecules caused the flow properties to change with time. It is proposed that hot film probes, positioned at different locations along the duct walls, be used to obtain these measurements. Since these probes are usually mounted flush with the walls, the disturbances (if any) introduced into a flow are expected to be minimal. Cross-correlations between instantaneous velocity and wall-shear stress will help provide further insight into the flow structure during drag reduction. Furthermore, secondary motions in ducts with non-circular cross-sections are known to be driven by gradients of the Reynolds normal and shear stresses. The Reynolds shear stress is greatly attenuated in highly drag reducing flows, but it is not clear whether there is a resulting change in the secondary flow pattern. To clarify this, SPIV measurements in a square duct can be conducted.

It is recommended that a similar procedure for obtaining the relationship between $\%DR$ and Wi in polyacrylamide solutions, in this thesis, be applied to other types of flexible polymers to test the universality of the observed correlation. An attempt to obtain data for polyethylene oxide in the experimental rigs was not successful, as the polymer degrades much quicker, resulting in large variation in rheological properties within each rig at the same pumping times. Unfortunately such inhomogeneity of degradation precluded the same approach as was possible for the polyacrylamides. However, in smaller or non-recirculating facilities, these issues may be less serious. Finally, it will be interesting to investigate the effect of polymer additives on heat and mass transport.

References

1. Uhlmann M, Pinelli A, Kawahara G, Sekimoto A (2007) Marginally turbulent flow in a square duct. J Fluid Mech 588:153–162
2. Biau D, Bottaro A (2009) An optimal path to transition in a duct. Phil Trans R Soc Lond A 367(1888):529–544
3. Shi L, Avila M, Hof B (2013) Scale invariance at the onset of turbulence in Couette flow. Phys Rev Lett 110(20):204502
4. Lundbladh A, Johansson AV (1991) Direct simulation of turbulent spots in plane Couette flow. J Fluid Mech 229:499–516
5. Tillmark N, Alfredsson PH (1992) Experiments on transition in plane Couette flow. J Fluid Mech 235:89–102
6. Daviaud F, Hegseth J, Bergé P (1992) Subcritical transition to turbulence in plane Couette flow. Phys Rev Lett 69:2511–2514
7. Gavrilakis S (1992) Numerical simulation of low-Reynolds-number turbulent flow through a straight square duct. J Fluid Mech 244:101–129
8. Nagata M (1990) Three-dimensional finite-amplitude solutions in plane Couette flow: bifurcation from infinity. J Fluid Mech 217:519–527
9. Faisst H, Eckhardt B (2003) Traveling waves in pipe flow. Phys Rev Lett 91(22):224502
10. Wedin H, Bottaro A, Nagata M (2009) Three-dimensional traveling waves in a square duct. Phys Rev E 79:065305
11. Uhlmann M, Kawahara G, Pinelli A (2010) Traveling-waves consistent with turbulence-driven secondary flow in a square duct. Phys Fluids 22(8):084102
12. Hof B, van Doorne CWH, Westerweel J, Nieuwstadt FTM, Faisst H, Eckhardt B, Wedin H, Kerswell RR, Waleffe F (2004) Experimental observation of nonlinear traveling waves in turbulent pipe flow. Science 305(5690):1594–1598
13. Hsu HW, Hsu JB, Lo W, Lin CA (2012) Large eddy simulations of turbulent Couette-Poiseuille and Couette flows inside a square duct. J Fluid Mech 702:89–101
14. Shahmardi A, Zade S, Ardekani MN, Poole RJ, Lundell F, Rosti ME, Brandt L (2019) Turbulent duct flow with polymers. J Fluid Mech 859:1057–1083

Appendix
Laminar Couette-Poiseuille Flows in a Square Duct

A.1 Introduction

In this chapter, the results of experiments and numerical simulations on laminar Couette-Poiseuille flows are presented. A new test-section for investigating these flows has been constructed. The goal is to allow for an investigation of the effect of wall motion on the turbulence field in a duct. Further details on the moving-wall test section are given in Sect. A.2. In designing such a facility, a knowledge of the flow development length (the distance from the inlet at which the velocity profile becomes invariant with streamwise distance) in both laminar and turbulent flow is very important, as this dictates how long the duct has to be for fully developed flow to be attained. Therefore, laminar numerical simulations for the determination of the entrance length were carried out during the design stage, both for the flow in a two-dimensional (2D) channel and also for the three dimensional (3D) flow in a square duct. Following the commissioning of the new rig, preliminary LDV measurements have been conducted in the laminar regime, for which analytical solutions to the fully developed flow are available for benchmarking purpose. At the Reynolds numbers considered, good agreement between experiment, numerical simulation and analytical solutions have been obtained. In the future, these experiments will be extended to turbulent flows.

A.2 Couette Flow Test Section

The test-section with a moving wall (see Fig. A.1) has a length of 1 m and was designed for use in the square duct rig. The side walls of the duct were left intact while the top wall was modified to allow for the introduction of wall translation. Translatory motion was implemented using an endless stainless steel belt (manufactured by Belt Technologies Europe) of width 100 and 0.127 mm thickness. The belt drive consists of three pulleys positioned as shown in Fig. A.1a; the system is driven by a motor

© Springer Nature Switzerland AG 2019
B. Owolabi, *Characterisation of Turbulent Duct Flows*,
Springer Theses, https://doi.org/10.1007/978-3-030-19745-2

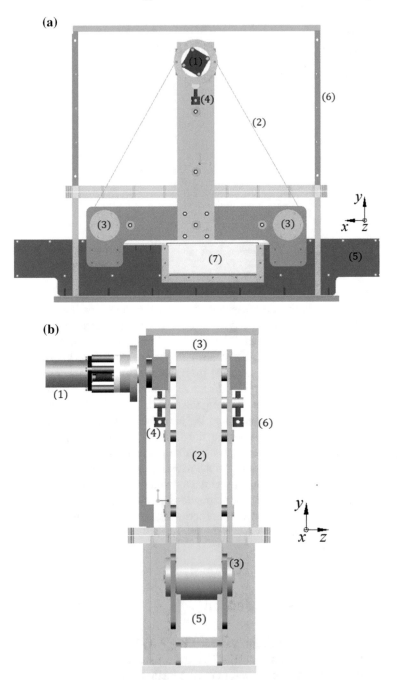

Fig. A.1 **a** Test section for the study of wall driven flow (**a**) side view: (1) electric motor; (2) stainless steel belt; (3) pulleys; (4) tracking screw; (5) square duct; (6) water-tight casing; (7) borosilicate glass window. **b** Test section for the study of wall driven flow (**b**) front view. Numbering as in (**a**)

attached to the top pulley. To guard against drifting of the belt, the top pulley is crowned and tracking screws are installed to allow for adjustment of the shaft. The drive is positioned such that the belt surface is flush with the top of the side walls (it should be noted that there will be a gap, however tiny, between the belt and the side walls; nevertheless, laminar flow measurements to be shown in Sect. A.5 exhibit good agreement with the analytical solution, indicating that the effect on the flow was negligible). This arrangement results in a moving wall of length 0.5 m corresponding to 12.5 h. LDV measurements in laminar Couette-Poiseuille flow at Re up to 67 showed that this length was enough to ensure a fully developed laminar flow without sagging of the belt. The maximum attainable wall speed is 5cm/s in both forward and reverse directions. The test section is enclosed in a casing made of PVC plastic, which contains the working fluid and glass windows are provided to allow for optical access to the flow.

A.3 Development Length in Duct Flows

Flow development and entrance effects in ducts have been the subject of extensive research, due to their huge importance in practical applications as well as in fundamental studies on the characteristics of duct flows [1]. From a prescribed velocity profile at the inlet, a flow field evolves in the streamwise direction, satisfying the no-slip condition at the walls. Boundary layers which grow with distance from the inlet are thus formed. Eventually, these boundary layers coalesce and the velocity profile becomes invariant with streamwise distance, the flow having become fully developed. The distance from the inlet at which this occurs is referred to as the development or entry length (L). The entrance region is usually associated with a higher pressure drop than in fully developed flow due to the increased shear stress at the wall.

Although the flow development phenomenon is well understood, it is surprising that correlations which accurately predict the entry length have only been developed fairly recently (see e.g. [2]). Using simple scaling arguments, [2] showed the important roles played by both molecular diffusion as well as convection in the flow development process. The former was ignored in most previous studies, thus leading to a wide variation in the predicted value of L. A detailed review on these studies can be found in [2]. For $Re \rightarrow 0$, where diffusion dominates, L/D (where D is the characteristic length scale of a duct, such as the pipe diameter or channel height) is essentially a constant, independent of the Reynolds number; however, at $Re \rightarrow \infty$, L/D varies linearly with Reynolds number, due to the dominance of convection. Hence an appropriate expression for the development length should be of the form

$$L/D = C_0 + C_1 Re. \tag{A.1}$$

In turbulent flows, the chaotic mixing by randomly fluctuating eddies causes the effect of molecular diffusion to be overshadowed, hence the entry lengths are much smaller than in laminar flow [3].[1]

For Laminar flow of a Newtonian fluid in pipes and two-dimensional channels, the following non-linear equations respectively hold [2]

$$L/D = [(0.619)^{1.6} + (0.0567\,Re_D)^{1.6}]^{1/1.6} \tag{A.2}$$

$$L/D = [(0.631)^{1.6} + (0.0442\,Re_D)^{1.6}]^{1/1.6}, \tag{A.3}$$

where $Re_D = \rho U_b D/\mu$. Similar correlations have also been proposed for concentric annuli [5] as well as non-Newtonian flows [6, 7]. These studies have considered only purely pressure-driven flows, hence wall-driven flows will be investigated in this chapter.

A.4 Numerical Simulations

Two sets of numerical simulations have been conducted. In the first, the flow in a 2D channel is computed. The problem is then extended to the 3D case (i.e. flow in a square duct). The governing transport equations are those expressing conservation of mass and momentum. For the 3D flow, the equations read

$$\frac{\partial u}{\partial x} + \frac{\partial v}{\partial y} + \frac{\partial w}{\partial z} = 0 \tag{A.4}$$

$$\rho\left(u\frac{\partial u}{\partial x} + v\frac{\partial u}{\partial y} + w\frac{\partial u}{\partial z}\right) = -\frac{\partial p}{\partial x} + \mu\left(\frac{\partial^2 u}{\partial x^2} + \frac{\partial^2 u}{\partial y^2} + \frac{\partial^2 u}{\partial z^2}\right)$$

$$\rho\left(u\frac{\partial v}{\partial x} + v\frac{\partial v}{\partial y} + w\frac{\partial v}{\partial z}\right) = -\frac{\partial p}{\partial y} + \mu\left(\frac{\partial^2 v}{\partial x^2} + \frac{\partial^2 v}{\partial y^2} + \frac{\partial^2 v}{\partial z^2}\right)$$

$$\rho\left(u\frac{\partial w}{\partial x} + v\frac{\partial w}{\partial y} + w\frac{\partial w}{\partial z}\right) = -\frac{\partial p}{\partial z} + \mu\left(\frac{\partial^2 w}{\partial x^2} + \frac{\partial^2 w}{\partial y^2} + \frac{\partial^2 w}{\partial z^2}\right) \tag{A.5}$$

---

[1] Most duct flows (such as those in a pipe or square duct) are linearly stable at all Reynolds numbers [4]; therefore a laminar flow can theoretically, be maintained indefinitely, hence L/D is expected to become very large as $Re \to \infty$. In practical applications, however, transition to turbulence occurs at much lower Reynolds numbers and the entry length is much lower than in laminar flow due to the diffusive nature of the turbulent eddies.

Equations A.4 and A.5 simplify to the following in the 2-D channel

$$\frac{\partial u}{\partial x} + \frac{\partial v}{\partial y} = 0 \qquad (A.6)$$

$$\rho\left(u\frac{\partial u}{\partial x} + v\frac{\partial u}{\partial y}\right) = -\frac{\partial p}{\partial x} + \mu\left(\frac{\partial^2 u}{\partial x^2} + \frac{\partial^2 u}{\partial y^2}\right)$$

$$\rho\left(u\frac{\partial v}{\partial x} + v\frac{\partial v}{\partial y}\right) = -\frac{\partial p}{\partial y} + \mu\left(\frac{\partial^2 v}{\partial x^2} + \frac{\partial^2 v}{\partial y^2}\right). \qquad (A.7)$$

In solving the above equations numerically, the commercial software, ANSYS Fluent, is utilised. This package has been extensively used in fluid dynamics research (see e.g. [6, 8, 9]). The momentum equations are discretised using the second-order upwind scheme, while coupling between velocity and pressure is implemented using the semi-implicit method for pressure-linked equations (SIMPLE, see [10]).

Figures A.2 and A.3 show the schematics of the computational domain for the numerical simulations and the co-ordinate systems employed. At the walls, the no slip boundary condition (i.e fluid velocity at the boundary is equal to the wall velocity) is imposed. Two types of inlet boundary conditions are examined. In the first, a uniform velocity profile (most commonly used in research on development length) is applied at the inlet. For this case, the entire topwall is made to translate (see Figs. A.2a and A.3a). In the second case, a parabolic velocity profile (corresponding to that of the fully-developed pressure-driven flow at a given bulk velocity) is introduced at the inlet (see Figs. A.2b and A.3b), and wall motion is imposed only after a streamwise distance of a few duct heights (D) from the inlet. This position is taken as the $x = 0$ location (in the previous case, $x = 0$ is located at the duct inlet). The wall motion is terminated at a few Ds from the outlet, hence the setup matches the configuration in the newly designed test-section for the square duct rig. The development length results presented in this chapter are based on the distance from the $x = 0$ location. In all cases, zero axial gradients are imposed at the outlet and the length of the moving-wall region was set to be a function of the Reynolds number, at least three times as long as the development length being computed. Computations on longer domains yielded identical results to those based on the above criterion, thus indicating that the solutions obtained are independent of the domain length.

The development length is defined as the axial distance from the $x = 0$ location, for the maximum velocity, u_{max}, to monotonically[2] attain a value which is within one percent[3] of its fully developed analytical value. For a purely pressure-driven flow developing from a uniform velocity profile at the inlet, u_{max} corresponds to the centreline velocity.

---

[2] An overshoot from the analytical solution can occur before the velocity at a wall-normal location approaches its fully-developed value (see [2, 7]).

[3] The exact fully developed value is approached asymptotically.

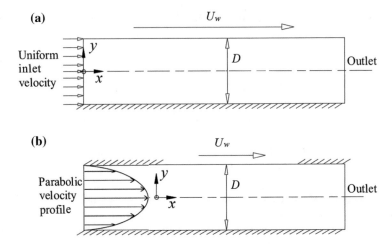

Fig. A.2 Computational domain and boundary conditions for the flow in a 2-D channel: **a** uniform inlet velocity, **b** parabolic velocity profile at the inlet

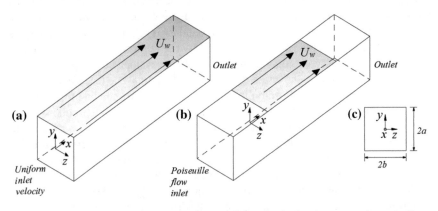

Fig. A.3 Computational domain and boundary conditions for the flow in a square duct: **a** uniform inlet velocity, **b** parabolic velocity profile at the inlet

However, in Couette-Poiseuille flows, the wall-normal location of u_{max} is shifted towards the moving wall with increasing values of the wall to bulk velocity ratio, U_w/U_b. In the limiting case of purely wall-driven flow, the peak velocity is located at the moving wall. The flow at this location instantaneously becomes fully developed as a result of the no-slip condition, hence in calculating the development length for this case, the entire velocity profile was examined (in the square duct, the velocity profiles along the vertical and horizontal wall bisectors were examined).

The laminar flow analytical solutions for 2D Couette-Poiseuille flow (see e.g. [11]) is given by

$$u = \frac{1}{2\mu}\frac{dp}{dx}\left[y^2 - h^2\right] + \frac{U_w}{2}\left[\frac{y}{h} + 1\right], \tag{A.8}$$

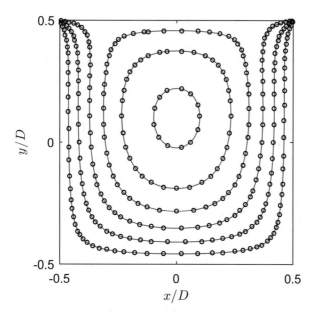

Fig. A.4 Comparison between u/U_b from analytical solution and numerical simulation at $U_w/U_b = 1$. Symbols represent numerical simulation while lines represent the analytical solution. Values of u/U_b shown, range from 0.3 to 1.8 with increment of 0.3. Translatory motion is introduced at the top wall

where h is the channel half height and the co-ordinate system is based on that of Fig. A.2. It can be shown that Eq. A.8 is simply a linear combination of the expressions for the purely pressure-driven and purely wall-driven flows. For a square duct, the Poiseuille flow analytical solution is (see [11])

$$u = U_b \frac{48}{\pi^3} \frac{\sum\limits_{i=1,3,5,\ldots}^{\infty} (-1)^{(i-1)/2} \left[1 - \frac{\cosh(i\pi y/2a)}{\cosh(i\pi b/2a)} \right] \frac{\cos(i\pi z/2a)}{i^3}}{1 - \frac{192a}{\pi^5 b} \sum\limits_{i=1,3,5,\ldots}^{\infty} \frac{\tanh(i\pi b/2a)}{i^5}}, \tag{A.9}$$

where a and b are as indicated in Fig. A.3. Similarly, for the purely wall-driven flow, the analytical solution is [12]

$$u = U_w \left(\frac{1-z}{2} + \sum\limits_{n=1}^{\infty} \frac{2(-1)^n}{n\pi} \cdot \sin \left[\frac{n\pi}{2}(1-z) \right] \cdot \frac{\cosh(n\pi y/2)}{\cosh(n\pi/2)} \right) \tag{A.10}$$

The Couette-Poiseuille equation is then a linear combination of Eqs. A.9 and A.10; simulation results from ANSYS Fluent are in excellent agreement with the resulting analytical solution for the fully-developed flow (see Fig. A.4).

A.4.1 Mesh Independence Studies

To select a suitable mesh for the numerical simulations, a series of computations were carried out at $Re_D = 0.5$, 1 and 100, using different grids ($Re_D = \rho U_b D/\mu$). For the 2D case, three structured cartesian meshes (M1, M2 and M3) were examined both for Poiseuille ($U_w/U_b = 0$) and Couette-Poiseuille flow with $U_w/U_b = 1$. The accuracy of the fully developed profiles obtained from the simulations was estimated from the relative error

$$E = \left| \frac{u_{max} - u_{max,a}}{u_{max,a}} \right| \times 100, \qquad (A.11)$$

where $u_{max,a}$ is the fully developed analytical value of the maximum streamwise velocity. The error (e) in the development length estimates (L_{Fluent}) from the meshes was also determined from

$$e = \left| \frac{L_{Fluent} - L_{extrap}}{L_{extrap}} \right| \times 100, \qquad (A.12)$$

where L_{extrap} is the development length obtained from a Richardson extrapolation of the results (see e.g. [10]), and provides an estimate of the development length which is more accurate than the solution from the finest grid, M3. By fitting the values of E to an equation of the form $E = (\Delta x)^p + c$, the accuracy, p, of the numerical scheme was estimated to be second order (i.e. $p = 2$). This value of p was used in computing the Richardson extrapolations.

Results of the grid independence study in the 2D channel are summarised in Table A.1. It can be observed that in meshes M2 and M3, the error values are small (E is less than 0.08%, while e is smaller than 0.7%). Since the difference in the value of L/D obtained from both meshes is less than 0.5%, all other simulations were conducted on grid M2.

For simulations in a square duct, six different meshes were tested (see Table A.2). In all the grids, E is less than 0.4%, indicating that the fully developed flow fields from the simulations are highly accurate. However, for grids M1, M2, M3 and M4, all uniform meshes, the error in the development length estimates are large ($e = 10.58\%$ in the finest uniform mesh, M4). In M4, the total number of cells required for the simulation at $Re_D = 1$ was about twelve million, hence the computational cost, in terms of memory requirement and simulation time was very high. Therefore, a further refinement was not feasible, given the limited computational resources available. Instead, non-uniform meshes, M5 and M6 were tested. These grids are symmetrically stretched in the cross-sectional plane of the duct, such that the size of the maximum grid cell at the centre is four times larger than the smallest grid at the wall. Uniform streamwise spacing is employed in the moving-wall section, with the number of cells given by $N_x = N_y.L_w/D$ (where L_w is the length of the moving wall). Twenty grid cells, each, were employed in the two regions with stationary top walls (see Fig. A.3b), the spacing increasing with distance from the moving wall. In mesh M6, e is less than 1.5%. Since the simulations at higher Reynolds numbers

Table A.1 Grid independence study for Couette-Poiseuille flow in a two-dimensional channel with uniform inlet velocity profile. N_y and N_x are the number of cells in the wall-normal and streamwise directions respectively, while L_w is the length of the moving wall

Mesh	$N_y \times N_x$	$\Delta x/D = \Delta y/D$	u_{max}/U_b	$E(\%)$	L/D	$e(\%)$
$U_w/U_b = 0$ (purely pressure-driven flow), $Re_D = 0.5$						
M1	$40 \times 40 L_w/D$	0.025	1.49496	0.274	0.650	2.52
M2	$80 \times 80 L_w/D$	0.0125	1.49953	0.072	0.638	0.63
M3	$160 \times 160 L_w/D$	0.00625	1.49988	0.019	0.635	0.16
Richardson extrapolation					0.634	
$U_w/U_b = 0$, $Re_D = 100$						
M1	$40 \times 40 L_w/D$	0.025	1.49792	0.139	4.865	3.53
M2	$80 \times 80 L_w/D$	0.0125	1.49942	0.039	4.723	0.51
M3	$160 \times 160 L_w/D$	0.00625	1.49984	0.011	4.705	0.13
Richardson extrapolation					4.699	
$U_w/U_b = 1$, $Re_D = 0.5$						
M1	$40 \times 40 L_w/D$	0.025	1.33208	0.132	0.877	1.50
M2	$80 \times 80 L_w/D$	0.0125	1.33302	0.034	0.867	0.35
M3	$160 \times 160 L_w/D$	0.00625	1.33325	0.006	0.865	0.12
Richardson extrapolation					0.864	

required longer domains, hence larger number of grid cells, this mesh was used for subsequent computations as it allowed for a good compromise between accuracy and computational cost. The DNS code described in Chap. 4 (which was run on a supercomputer, thus allowing for more grid cells) required a periodic boundary condition (which enforced a fully developed flow across the entire domain), hence it was not suitable for development length computations.

A.4.2 Development Length Computations

Figure A.5 shows the variation of development length with Reynolds number for Couette-Poiseuille flows in a 2D channel, with uniform inlet velocity profiles. The results cover U_w/U_b values spanning from zero (purely pressure-driven flow) to two[4] (purely wall-driven flow).

For the purely pressure-driven flow, the results are in good agreement with the correlation of [2]. As U_w/U_b is increased, the development lengths can be observed to become higher. The implication of this is that longer ducts are required for studies on fully developed wall-driven flows. In the creeping flow regime ($Re_D \to 0$), where diffusion dominates, L/D is essentially constant, and its value is of the order of the

---

[4]It can be shown that $U_b = U_w/2$ in purely wall-driven flow; therefore, $U_w/U_b = 2$.

Table A.2 Grid independence study for Couette-Poiseuille flow in a square duct at $Re_D = 1$ and $U_w/U_b = 2/3$. Inlet velocity profile is parabolic

Mesh	$N_y \times N_x$	$\Delta y/D =$ $\Delta z/D$	u_{max}/U_b	$E(\%)$	L/D	$e(\%)$
Uniform mesh						
M1	$50 \times 50 \times$ $50L_w/D$	0.0200	1.92086	0.370	0.84	55.84
M2	$60 \times 60 \times$ $60L_w/D$	0.0167	1.92266	0.277	0.783	45.27
M3	$100 \times 100 \times$ $100L_w/D$	0.0100	1.92626	0.090	0.62	15.03
M4	$120 \times 120 \times$ $120L_w/D$	0.0083	1.92630	0.088	0.596	10.58
Richardson extrapolation					0.539	
Non-uniform mesh						
M5	$90 \times 90 \times$ $(90L_w/D + 40)$		1.92446	0.184	0.556	3.15
M6	$95 \times 95 \times$ $(95L_w/D + 40)$		1.92582	0.114	0.547	1.48

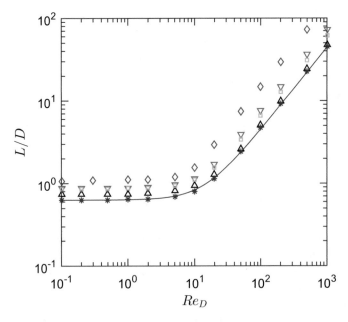

Fig. A.5 Development lengths in a 2D channel with uniform inlet velocity, at different wall to bulk velocity ratios. *, $U_w/U_b = 0$ (Poiseuille flow); $\triangle$, $U_w/U_b = 2/3$; $\square$, $U_w/U_b = 1$; $\triangledown$, $U_w/U_b = 4/3$; $\diamond$, $U_w/U_b = 2$ (purely wall-driven flow); ——, correlation of [2]

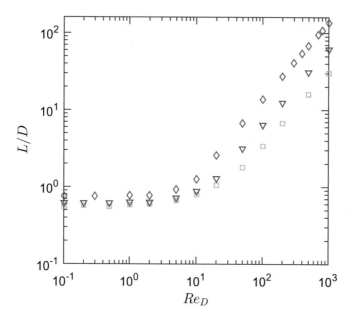

Fig. A.6 Development lengths in a 2D channel with parabolic inlet velocity, at different wall to bulk velocity ratios. $\square$, $U_w/U_b = 1$; $\triangledown$, $U_w/U_b = 4/3$; $\diamond$, $U_w/U_b = 2$ (purely wall-driven flow)

channel diameter, as expected. For $Re_D > 50$, L/D increases linearly with Re_D, while a non-linear behaviour can be observed in the region $2 < Re < 50$, consistent with the findings of [2, 5, 6].

A similar trend can be observed in the simulations with parabolic inlet velocity profiles (see Fig. A.6). However, the flow develops quicker, thus showing the dependence of L/D on the inlet conditions. Since in the problem setup, the moving wall is situated between two stationary walls (see Fig. A.3), a further developing region is expected before the end of the translating wall is reached. Here, the velocity profile gradually changes to match the conditions in the stationary-wall region downstream. As this requires information to be propagated in the upstream direction, the process is dominated by molecular diffusion, and as expected, this development length was found to be of order D.

An attempt to collapse the data for different U_w/U_b onto a single plot was unsuccessful. The only length scale present in the problem is the channel diameter. Plotting L/h against Re_D merely caused the entire data to be scaled by a factor of two in the vertical axis. Similarly, expressing the Reynolds number in terms of u_{max} or centreline velocity, did not yield a collapse in the data. This is also the case for the 3D results which will be discussed in the following paragraphs.

In the purely pressure-driven flow in a square duct with a uniform inlet velocity profile, the development length is greater than in a 2D channel (see Fig. A.7, where the correlation of [2] for channel and pipe flow are included), hence 3D effects introduced by the side walls cannot be neglected. The L/D estimates in the square duct are much

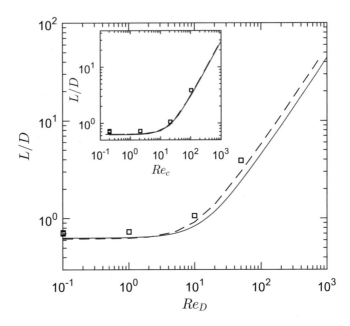

Fig. A.7 Development lengths in purely pressure-driven flows. □, square duct simulations. Dashed and solid lines represent the correlations of [2] for laminar flow in a pipe and 2D channel respectively. Inset shows plots where Reynolds number is expressed in terms of the centreline velocity, U_c (i.e. $Re_c = \rho U_c D / \mu$)

closer to those in the pipe. A possible reason for the observed difference is the non-uniform distribution of the wall shear stress in the former, especially, close to the corners. It is therefore expected that similar results will be obtained in both geometries if the square duct corners are rounded. A better collapse in the correlations of [2] for pipes and 2D channels can be observed when the Reynolds number is expressed in terms of the centreline velocity (see inset of Fig. A.7), but L/D is still slightly higher in the square duct (with a maximum difference of about 20%).

In square duct Couette-Poiseuille flows, the development length can be observed to become larger with increasing U_w/U_b, similar to the observations in the 2D channel (see Fig. A.8). As a result of the computational difficulties earlier pointed out, only a limited number of simulations were conducted at $Re_D > 10$. Figure A.9 shows the variation of L with U_w/U_b in the limit as $Re_D \rightarrow 0$. As expected, L is higher in ducts with uniform inlet velocity profiles. The data is fairly well fitted ($R^2 = 0.985$) by a linear curve: $L/D = C_0 = 0.1366 U_w/U_b + 0.7673$. For the simulations with parabolic inlet velocities, a line with the same slope, but a smaller intercept (0.49), fits well to the data in the region $0.5 < U_w/U_b < 4$. It should be noted that the development length in the limiting case of purely pressure-driven flow (i.e. $U_w/U_b = 0$) must be equal to zero, as a result of the inlet boundary condition, hence it is not surprising that there is a non-linear behaviour in the region, $U_w/U_b < 0.5$. In Fig. A.8, the data points at $Re_D > 10$ appear to diverge after being roughly parallel

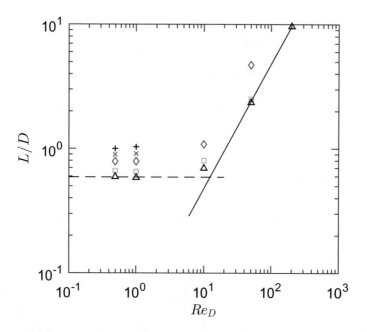

Fig. A.8 Couette-Poiseuille flow development length simulations in a square duct. $\triangle$, $U_w/U_b = 2/3$; $\square$, $U_w/U_b = 1$; $\diamond$, $U_w/U_b = 2$; $\times$, $U_w/U_b = 3$; $+$, $U_w/U_b = 4$. - - - -, $L/D = 0.59$; ———, $L/D = 0.0477 Re_D$

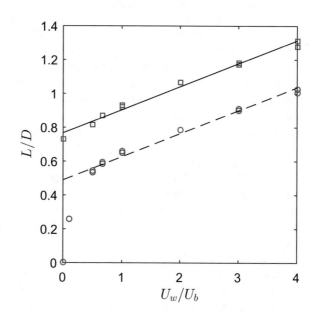

Fig. A.9 Development length at $Re_D \to 0$. $\square$, uniform inlet velocity profile; $\bigcirc$, parabolic inlet velocity profile. Blue symbols: $Re_D = 0.5$; red symbols: $Re_D = 1$. ———: $L/D = 0.1366(U_w/U_b) + 0.7673$, $[R^2 = 0.985]$; - - - -: $L/D = 0.1366(U_w/U_b) + 0.49$

in the creeping flow regime, hence values of the slope, C_1, in the $Re_D \to \infty$ limit cannot be inferred from the plots in Fig. A.9. For $U_w/U_b = 2/3$, where numerical data up to $Re_D = 200$ has been obtained, a linear fit to the data points in Fig. A.8 indicates that $C_1 \approx 0.0477$ in the high Reynolds number limit.

From the 2D and 3D simulations, it can be inferred that fully-developed laminar Couette-Poiseuille flows can be obtained in the new test section of length $6.25D$ up to $Re_D \approx 100$.

A.5   Experiments

LDV measurements in laminar Couette-Poiseuille flows have been conducted to investigate the flow development and also to determine how accurately the analytical solution can be reproduced in the new moving-wall test section of the square duct experimental rig. The Poiseuille component was generated by pumping fluid through the duct at different flow rates. To these pressure-driven flows, a Couette component was introduced by translating the stainless steel belt. In all the measurements, the wall moves in the same direction as the bulk flow and Glycerol/water solution (70/30% by volume) with an Ekman number of about 12 was employed as working fluid. The results of laminar flow simulations at $Ek = 12$ (Coriolis body force included) are identical to those in which no body forces are included; hence Coriolis effects (see [13]) are negligible at this Ekman number.

Figure A.10a and b show the velocity data at different distances from the inlet in Couette-Poiseuille flows with $U_w/U_b = 1.7$ and 1. The measurement locations ($y/D = 0.3$ and 1, respectively) correspond to positions of maximum streamwise velocity in the analytical solution. The uncertainty in the LDV measurements is estimated to be about 3% (as indicated by the error bars), hence, an accurate determination of the development length is nearly impossible, since the criteria used requires that the velocity profiles be within 1% of the laminar analytical solution. This explains why an accurate experimental determination of L have so far been unsuccessful (see [2]). In Fig. A.10a ($U_w/U_b \approx 1.7$), an increase in the measured velocity with increasing streamwise distance can be observed, with a plateau region occurring at about $x/D \gtrsim 4$. However, in all but the first data point, the analytical solution is within the error bounds of the measurements. For $U_w/U_b \approx 1$ (Fig. A.10b), it appears that the flow is already fully-developed at the first measurement location accessible by the LDV probe ($x/D \approx 2.35$). This is consistent with the simulation results (see Fig. A.8).

From the afore-reported measurements, it is clear that at the Reynolds numbers considered ($Re_D \approx 63$ and $Re_D \approx 76$ for $U_w/U_b \approx 1.7$ and 1, respectively), fully-developed flows can be obtained at $x/D \approx 4.69$ (the furthest downstream position accessible by the LDV probe). Figure A.11 shows the velocity profiles at this location, for $U_w/U_b \approx 1.7$. A good agreement with the analytical solution can be observed at this bulk to wall-velocity ratio and so is the case with $U_w/U_b \approx 1$ (see Fig. A.12). Close to the moving wall ($y/D = 0.375$, see Fig. A.11b) the velocity profile from

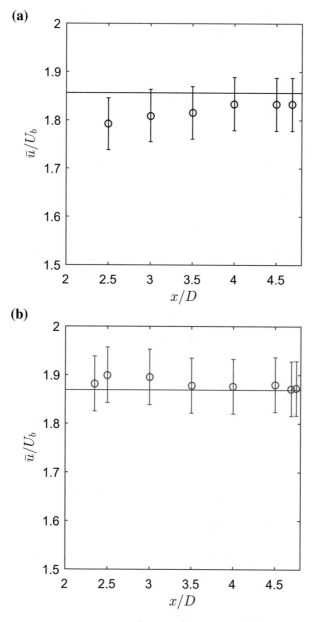

Fig. A.10 Couette-Poiseuille flow development: **a** $U_w/U_b \approx 1.7$ and $Re_D \approx 63$; **b** $U_w/U_b = 1$ and $Re_D \approx 76$. Measurements have been taken at $y/D = 0.3$ and $y/D = 0.1$ along the vertical wall bisector, corresponding to the respective maximum velocity locations. Solid lines are the laminar flow analytical solutions. Error bars representing $\pm 3\%$ measurement uncertainty have been included

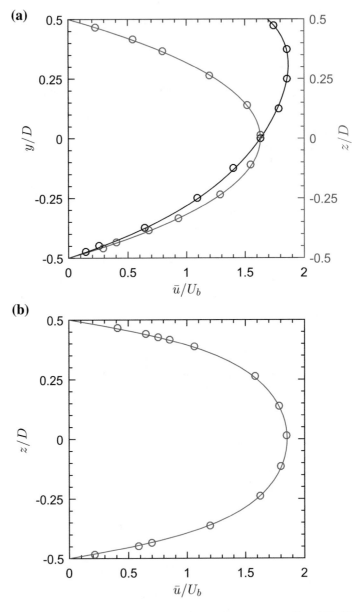

Fig. A.11 Velocity profiles in laminar Couette-Poiseuille flow at $Re_D \approx 63$ and $U_w/U_b \approx 1.7$: **a** fully developed profiles at $x/D \approx 4.69$; black lines/symbols are the profiles along the vertical wall bisector (from the bottom stationary wall to the top moving wall), while red lines/symbols are those along the horizontal wall bisector (from one stationary side wall to the other), **b** fully developed profile at $y/D = 0.375$ (10 mm from moving wall, where measurements have been taken from one stationary side wall to the other at $x/D \approx 4.69$). Solid lines represent the laminar flow analytical solution, while symbols represent LDV measurements

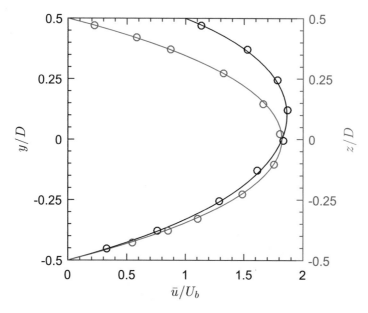

Fig. A.12 Fully-developed velocity profile in laminar Couette-Poiseuille flow at $U_w/U_b \approx 1$ and $Re_D \approx 76$. Solid lines represent the laminar flow analytical solution, while symbols represent LDV measurements. Black lines/symbols are the profiles along the vertical wall bisector (from the bottom stationary wall to the top moving wall), while red lines/symbols are those along the horizontal wall bisector (from one stationary side wall to the other)

side to side can be observed to agree well with the analytical solution. As indicated in Chap. 3, a finite gap must exist between the moving belt and the side walls; it is therefore reassuring to find out that this has a negligible effect on the flow at least at the low Reynolds numbers considered.

Velocity measurements in the developing flow at $x/D \approx 2.5$ for $U_w/U_b \approx 1.7$ are shown in Fig. A.13. A parabolic profile, different from the analytical solution, is obtained along the horizontal wall bisector; the profile along the vertical wall bisector is also nearly uniform across the top half of the duct. The results agree well with numerical simulation data from ANSYS Fluent. From the results so far, it is clear that good quality data on Couette-Poiseuille flows can be obtained from the newly designed test section at least in laminar flow.

A.6 Summary

A test-section with a moving wall has been designed and constructed for the square duct experimental rig, to facilitate a study on the effect of wall motion on the turbulence field in duct flows. Prior to performing these turbulent flow experiments,

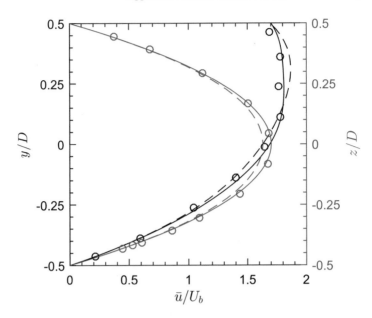

Fig. A.13 Velocity profiles in developing flow at $x/D \approx 2.5$. Solid lines represent the numerical simulation results from ANSYS Fluent, while dashed line are the analytical solutions for fully developed flow, and symbols represent LDV measurements. Black lines/symbols are the profiles along the vertical wall bisector (from the bottom stationary wall to the top moving wall), while red lines/symbols are those along the horizontal wall bisector (from one stationary side wall to the other)

it is important to investigate the flow in the laminar regime, for which analytical solutions are available for benchmarking purposes. A knowledge of the flow development length is also very crucial, given the finite size of the experimental facility. In this regard, numerical simulations have been conducted, first, for the flow in a 2D channel, with an extension to the 3D case of flow in a square duct.

The results show that the development lengths (L) in Couette-Poiseuille flows are higher than those in purely pressure-driven flows, L becoming larger as the wall to bulk velocity ratio is increased. An attempt to collapse the data onto a single curve, using various velocity and length scales, was not successful. Consistent with the findings of [2], L was observed to be essentially constant, and of the order of the channel diameter, D, in the creeping flow regime (i.e. as $Re_D \rightarrow 0$). However, for $Re_D > 50$, the development length increased linearly with Reynolds number. The inlet boundary condition was found to be very important, as L was found to be higher in flows with uniform velocity profiles at the inlet than those with parabolic velocity profiles. LDV measurements in the new test section indicate that fully developed laminar Couette-Poiseuille flows can be obtained at the Reynolds numbers tested. The experimental data in the fully-developed region at $U_w/U_b = 1.7$ and 1 agree well with the laminar analytical solution. Measurements in the developing region are also in good agreement with the numerical simulation results.

Having tested the new rig in the laminar regime, the next step will be to investigate turbulent flows. Further details on such proposed future work in the new test section are given in Chap. 8.

References

1. McComas S (1967) Hydrodynamic entrance lengths for ducts of arbitrary cross section. ASME J Basic Eng 89(4):847–850
2. Durst F, Ray S, Ünsal B, Bayoumi O (2005) The development lengths of laminar pipe and channel flows. ASME J Fluids Eng 127(6):1154–1160
3. Yunus AC, Cimbala JM (2006) Fluid mechanics fundamentals and applications. McGraw-Hill
4. Tatsumi T, Yoshimura T (1990) Stability of the laminar flow in a rectangular duct. J Fluid Mech 212:437–449
5. Poole RJ (2010) Development-length requirements for fully developed laminar flow in concentric annuli. ASME J Fluids Eng 132(6):064501
6. Poole RJ, Ridley BS (2007) Development-length requirements for fully developed laminar pipe flow of inelastic non-Newtonian liquids. ASME J Fluids Eng 129(10):1281–1287
7. Poole RJ, Chhabra RP (2010) Development length requirements for fully developed laminar pipe flow of yield stress fluids. ASME J Fluids Eng 132(3):034501
8. Dennis DJC, Seraudie C, Poole RJ (2014) Controlling vortex breakdown in swirling pipe flows: experiments and simulations. Phys Fluids 26(5):053602
9. Escudier MP, O'Leary J, Poole RJ (2007) Flow produced in a conical container by a rotating endwall. Int J Heat Fluid Flow 28(6):1418–1428
10. Ferziger JH, Peric M (2012) Computational methods for fluid dynamics, 3rd edn. Springer Science & Business Media, London
11. White FM (2006) Viscous fluid flow, 3rd edn. McGraw-Hill, New York
12. Davoodi M, Lerouge S, Norouzi M, Poole RJ (2018) Secondary flows due to finite aspect ratio in inertialess viscoelastic Taylor-Couette flow. J Fluid Mech 857:823–850
13. Draad A, Nieuwstadt F (1998) The earth's rotation and laminar pipe flow. J Fluid Mech 361:297–308

Printed in the United States
By Bookmasters